T0291637

AN INTRODUCTION TO NUCLEAR PHYSICS

AN INTRODUCTION TO NUCLEAR PHYSICS

BY

N. FEATHER, F.R.S.
Professor of Natural Philosophy in the University of Edinburgh

CAMBRIDGE
AT THE UNIVERSITY PRESS
1948

CAMBRIDGE
UNIVERSITY PRESS

University Printing House, Cambridge CB2 8BS, United Kingdom

Cambridge University Press is part of the University of Cambridge.

It furthers the University's mission by disseminating knowledge in the pursuit of education, learning and research at the highest international levels of excellence.

www.cambridge.org
Information on this title: www.cambridge.org/9781316509678

First published 1948
First paperback edition 2015

A catalogue record for this publication is available from the British Library

ISBN 978-1-316-50967-8 Paperback

CONTENTS

PART III. CONCERNING UNSTABLE NUCLEI

PART IV. TRANSFORMATIONS PRODUCED BY FAST-MOVING PARTICLES AND BY RADIATION

LIST OF PLATES

PREFACE

If the naive distinction between experimental and theoretical branches of a science be maintained, the present book must clearly be regarded as dealing with the experimental side of the science of nuclear physics. On the other hand, the farther removed the subject matter of any science is from common experience, the more trivial the distinction between "experimental" and "theoretical" becomes, in application to that science. It is certainly a trivial distinction in respect of nuclear physics. Individual researches may be carried through by "experimenters" and "theorists", respectively; they become significant, however, only when their results are fused in a common statement. This is the point of view from which succeeding pages have been written: it will probably appear most plainly in Part I, where, for the most part, well-known results are under discussion.

In Part I the attempt is made to trace the growth of the necessity of the concepts "nuclear atom" and "atomic-nucleus-possessing-internal-structure" for the progress of research in physics; Parts II, III and IV summarise the developments of the subject which followed upon the acceptance of this general scheme of interpretation. Again, the presentation may be said to be chiefly experimental in these later parts, if only for the reason that a change in the conceptual scheme would call for less alteration here than in any parallel, but more formal, treatment. Phenomena, rather than the mental constructs in terms of which they are variously interpreted—according to the accident of the time—are really fundamental. For this reason all discussion has been kept as far as possible phenomenological.

The book, then, is an introduction and a summary: the chief ideas necessary for an understanding of current research in nuclear physics have been given with the aid of a few illustrative examples; on the other hand all the important results have also been included, in tabular form.

Finally, a number of acknowledgments remains to be made: to Dr R. R. Nimmo who has read through the whole of the proofs with characteristic energy, to Dr H. S. W. Massey who has offered the opinion of a theorist on many matters under dispute, to Professor H. J. Taylor for the originals of Plates I (*b*) and III (*e*) and (*f*), to Dr T. W. Bonner for the original of Plate II (*d*) and to Dr G. Gamow an acknowledgment of a less precise nature. During the preparation of Dr Gamow's forthcoming *Structure of Atomic Nuclei and Nuclear Transformations* the task of revising the manuscript for publication was allotted to me. I wish to acknowledge the many advantages which the performance of this task conferred; I trust they have not here been abused, in any particular. Presently, Dr Gamow's book and the extensive monograph of Drs Bethe and Bacher[a] will provide all the "theoretical" background that any reader of these pages is likely to require.

NORMAN FEATHER

LIVERPOOL
3 *September* 1936

NOTE TO SECOND IMPRESSION

Pending the preparation of a new edition this photographic reprint of the first edition is issued to meet the urgent needs of University students.

N.F.

1948

[a] Part A, *Rev. Mod. Phys.* 8, 82, 1936; Part B, in preparation.

CHAPTER I

EXPERIMENTS—AND THE TYPES OF INFORMATION OBTAINED

§ 1. *Introduction: calculated and measured magnitudes.* The atomic theory of matter, as an element in speculative philosophy, is of considerable antiquity; as a scientific formulation it is to be attributed in the first place to Dalton.[a] From a consideration of the relative combining weights of elementary substances and the pressure-volume-temperature relations for isolated portions of matter in the gaseous state,[b] Dalton was led to the idea of atoms, identical amongst themselves when a single chemical substance is in question, but differing in respect of weight from one substance to another. Similarly, macroscopic phenomena, in particular capillarity and cohesion in liquids, provided the data which led Young[c] to an early and reasonable estimate of the "size" of an atom. Later Kelvin[d] made several such estimates on the basis of similar material. In each case the crude experimental data consisted of mass and length determinations: masses of chemical substances, linear dimensions of glass vessels, lengths of columns of mercury or of pointer movements over fixed scales. Transition to the sub-microscopic took place on paper—in the course of the arithmetic—or in the mind of the interpreter. The concept atom-of-matter arose in this way, and the necessary consequence of accepting such a concept—the fact that a certain degree of spatial extension had to be assigned to the atom—received numerical expression merely as a result of the calculations. The further history of this concept is the story of modern physics in one of its major aspects. As that history is briefly sketched in the sections

[a] Dalton, *New System of Chemical Philosophy*, Part I, 1808; Part II, 1810.

[b] Roscoe and Harden, *New View of the Origin of Dalton's Atomic Theory*, 1896.

[c] Young, "Cohesion of fluids", *Encyclopaedia Britannica*, 5th edition, 1816; *Collected Works*, 1, 461, 1855.

[d] Kelvin, *Baltimore Lectures*, pp. 279–84, 1904.

which follow it will be seen how the notion of spatial extent is particularised and the idea of structure is introduced, under the necessity of interpretation for more detailed experiments. Moreover, the main concern of the present book begins only after the broad outlines of this structural pattern have been accepted and it becomes necessary to particularise still further and assign structure to the atomic nucleus. Parts II, III and IV deal exclusively with the experiments which have made necessary these more recondite developments in the interpretative scheme.

At the very beginning, however, it is worth while insisting that the crude data of any scientific experiment, whatever its aim, must be measured physical quantities of roughly a single order of magnitude. Apparatus is designed to bring within compass of human perception and measurement significant evidence of events of the astronomical or the subatomic order; in the quantitative interpretation of these results powers of ten arise in the course of arithmetical multiplication and division—sometimes with alarming facility. Only the engineer is left in a position to evaluate by direct apprehension the import of his conclusions. The astronomer and the modern physicist, at opposite extremes of the scale, almost naturally discard inconvenient powers of ten for ease in forming such mental pictures as are required for successful thinking. But they invite danger whenever they forget what they have discarded. To such danger the nuclear physicist is particularly liable; he measures in terms of centimetre scales and gram "weights" and is just as far from quantities of this order of magnitude in stating the mass of the proton, for example, as the astronomer is in giving the distance of the spiral nebulae—the atoms in one gram of hydrogen in line at one centimetre intervals would stretch for nearly a million light years! It is the object of this chapter, then, to describe precisely the types of measurement made in experiments on nuclear physics, as it will be of succeeding chapters of Part I to enquire how far the interpretation of such experiments may be carried in terms of the concepts of macroscopic physics—the ideas of the engineer and the astronomer—or to what extent an entirely new point of view may be necessary. This new viewpoint is embodied in the wave mechanics of the theorist.

For convenience in description, and also because some approximation to chronological order is thereby attained, the experiments in question may be classified roughly as follows: (*a*) experiments concerning the properties of radioactive material or of matter subjected to the action of radiations from radioactive substances, (*b*) investigations of matter in general by the method of mass spectroscopy, (*c*) investigations of matter by the method of optical spectroscopy, and (*d*) experiments involving the bombardment of matter by artificially accelerated particles.

§ 2. *Experimental methods in radioactivity.* Radioactivity is a spontaneous activity of matter, first noticed in 1896,[a] which is confined, in general experience, to a relatively small number of elementary substances of high atomic weight. This activity is shown in a persistent, if gradually dwindling, emission of energy[b] which may be set in evidence in a number of ways. Thus the active material and portions of surrounding matter increase in temperature if thermally insulated from other bodies in the neighbourhood, volumes of gas previously possessing very feeble electrical conductivity become conducting, certain phosphorescent substances emit light and unexposed photographic plates become developable under the action of the radiations. Closer analysis discloses further the mechanism of these effects.

Conductivity, for example, may first be studied as a volume effect, the current through the gas being measured by the rate of charge of the quadrants of a sensitive electrometer, or by some similar arrangement. It is thus investigated in its dependence on the collecting field, on the distance of the preparation, the amount of screening matter interposed, or on the time. From another angle of approach it appears that the processes of expenditure of energy respon-

[a] Becquerel, *Comptes rendus*, 122, 420, 501, 1896.

[b] "Radiation" is the term generally applied to the vehicle by which energy is given out by matter.

sible for the conductivity are localised in the gas along linear elements; in the finer analysis, therefore, these processes do not occur uniformly throughout the volume. The development of the expansion chamber by C. T. R. Wilson[a] made this analysis possible. Here the sample of gas under observation is rendered temporarily supersaturated with water vapour and it is found that, in circumstances wherein an exactly similar sample of dry gas would show electrical conductivity, the volume of moist gas exhibits the condensation of vapour in "tracks". There is every reason to conclude[b] that these tracks represent the initial distribution in the gas of carriers of electricity such as are responsible for the conductivity. The distribution itself is most simply explained if it be assumed that the tracks delineate precisely the geometrical paths of single particles, most of which are emitted directly from the source, although some clearly are liberated from the gas in the chamber. By an obvious application of terms the particles emitted from the source are referred to as primary[c] particles, those set free in the body of the gas as secondary particles. All these particles lose kinetic energy in passing through the gas. Ionisation, that is the separation of opposite charges initially combined in the form of electrically neutral gas molecules, would obviously be impossible without some such expenditure of energy. Qualitatively, individual tracks show great differences in respect of density, length and degree of straightness (Plate I(*a*)). If the explanation in terms of ionising particles be adopted, at least two different types of primary particle must be postulated. This is a conclusion which, in actual fact, had been reached long before the expansion method was applied to the problem. Rutherford[d] had shown that some such assumption was necessary to explain the dependence of the saturation current through a given volume of gas on the amount of screening matter inter-

[a] Wilson (C. T. R.), *Phil. Trans. Roy. Soc.* 189, 265, 1897.

[b] Wilson (C. T. R.), *Proc. Roy. Soc.* 85, 285, 1911.

[c] This does not represent strictly the current usage of the term "primary β particle", however further refinement is impossible at the present stage (see footnote, p. 38).

[d] Rutherford, *Phil. Mag.* 47, 109, 1899.

Plate I

a

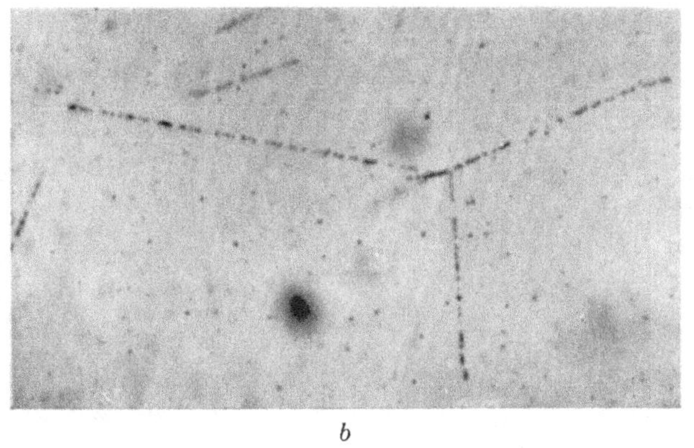

b

posed between the radioactive preparation and the gas.[a] In this way he had distinguished between "α" radiation, the effect of which—the major effect of the source—was cut off by a small amount of matter, and "β" radiation, which could be detected through much greater thicknesses. The investigation was carried farther by Bragg[b] who measured the saturation current between two parallel metal grids separated by a small distance in air, as a function of the distance of the source. He showed that the ionising effect of the α radiation ceased abruptly at a particular distance. In this way the idea of α particle "range" arose—some years before expansion chamber photographs provided direct evidence in the form of straight dense tracks having well-defined end points. To this early period, also, belong certain other experiments of Rutherford.[c] Carrying the absorption method farther than previous observers had done, he found a small residual effect through thick screens which was but slowly reduced as the thickness of absorber was still further increased. Villard[d] had already had photographic evidence of this in 1900. Both effects are to be attributed to a third type of primary radiation ("γ" radiation). Later researches have shown that the secondary ionising particles set in evidence by expansion chamber photographs result from the action of this radiation on the material through which it passes. These photographs illustrate clearly the reason for the great penetrating power of the γ radiation: the individual processes by which it loses energy occur at relatively long intervals in the course of its propagation through matter. It is sometimes convenient to refer to it as a non-ionising radiation, however obvious it is that such a description is not entirely exact.

Closer analysis of the action of the radiations on phosphorescent materials in many cases shows only that the phenomena involved are of considerable complexity, but with

[a] The experimental method here exemplified is generally referred to as the "absorption" method, a graph of ionisation current against amount of interposed matter being termed an absorption curve.

[b] Bragg and Kleeman, *Phil. Mag.* 8, 726, 1904.

[c] Rutherford, *Phys. Z.* 3, 517, 1902; *Nature*, 66, 318, 1902.

[d] Villard, *Comptes rendus*, 130, 1010, 1178, 1900.

one or two substances it has resulted in the discovery of an effect which has been put to much practical use. Crookes[a] showed that the general luminosity of phosphorescent zinc sulphide under the impact of the radiations consisted of individual flashes of light separately visible to the dark-adapted eye, when suitably aided. Simplest conditions obtain when the phosphorescent material is placed within the range of the α particles from a weak source. Then the flashes of light (scintillations) may be regarded as due to the impact of individual α particles on the screen. In this process a portion of the energy of the particle, having first been used, presumably, in ionisation and excitation of the atoms of solid matter traversed, is eventually radiated as visible light. Regener[b] first made use of this effect to register the arrival of α particles on a screen of determined area; in applying it in this way he has been followed by almost all the workers in radioactivity of the period 1909–29. During the last few years, however, equivalent electrical methods have been developed, having much greater power. These permit of the continuous registration of the passage of single particles, by amplification of their ionising effects. What is finally observed is no longer a small portion of the energy of the particle, as in the scintillation method, but energy from an external supply released by a type of trigger action controlled by the production of ionisation by the particle in its passage through the counting chamber. Thus a considerable deflection of a spot of light on a scale, or a loud sound in a telephone receiver, may result. In this connection two types of amplification must be distinguished. The distribution of applied electric field in the counting chamber may be such that the cumulative process of ionisation by collision occurs in the gas, so that considerably more charge is collected than is liberated by the passage of the particle itself—in this case a relatively small degree of further amplification by the electrical recording system is sufficient—or the initial ionisation may be simply collected by means of a smaller field, when the whole

[a] Crookes, *Proc. Roy. Soc.* 71, 405, 1903.
[b] Regener, *Verh. d. D. Phys. Ges.* 10, 78, 1908.

of the necessary amplification must be obtained from the electrical circuits employed.[a] Each arrangement has its own advantages and disadvantages. Smaller initial amounts of ionisation may be detected with the former arrangement ("point-counter", etc.), but the latter (ionisation chamber and valve amplifier) is more easily made "proportional", that is, arranged to give a final effect which provides a direct measure of the amount of charge liberated by the particle. As applied to experiments in radioactivity the point counter (or tube counter) is more suited to the investigation of β particle effects, the ionisation chamber and amplifier to phenomena in which α particles are involved.

Photography is an important aid both to electrical counting methods and also when the expansion chamber is employed. In the case of the expansion chamber it is obvious that this must be so; in the other case it is frequently of interest to know not only the number of particles registered within a given time, but also, as is possible with proportional amplifiers, the amount of ionisation which each particle has produced. In such circumstances photographic registration of the oscillograph deflections becomes a great convenience. But it has also been mentioned that radioactive radiations exert a direct action on the sensitive grains in a photographic emulsion. This fact is responsible for two methods of employing photographic materials as primary detectors in certain investigations. In the first method the radioactive matter is incorporated in the emulsion or is brought in contact with it. Then, on development, short rows of reduced silver grains appear[b] which must be interpreted as essentially α particle tracks (see Plate I(*b*)). The method is extremely crude, in that the tracks are short (about 5×10^{-3} cm.) and never contain very many grains (40, at most, with specially prepared emulsions), but it has the convenience for some purposes that the final image gives the integrated effect of

[a] Greinacher, *Z. Physik*, 36, 364, 1926; *ibid.* 44, 319, 1927; Wynn-Williams, *Proc. Camb. Phil. Soc.* 23, 811, 1927; Wynn-Williams and Ward, *Proc. Roy. Soc.* 131, 391, 1931.

[b] Kinoshita, *Proc. Roy. Soc.* 83, 432, 1910; Reinganum, *Verh. d. D. Phys. Ges.* 13, 848, 1911.

the radiation, which may be too feeble to be detected by other means.[a] On the other hand it is not applicable to the study of the individual particles constituting radioactive β radiation. The second method of employing photographic plates directly is to use them as detectors when canalised beams of particles are under investigation in vacuo—for they may be introduced into evacuated vessels without finally impairing the "vacuum". The deflection of such a beam of particles under the action of electric or magnetic fields may thus be studied in terms of the movement of the "image" formed on the plate by the normal incidence of the beam. This method will be further discussed in the section on mass spectrum analysis.

These methods then, simple measurement of ionisation current[b] (or of heating effect—which is occasionally used), counting of scintillations, direct exposure of the photographic plate, the method of the expansion chamber and the various arrangements of counter and proportional ionisation amplifier, enable a detailed study to be made of the properties of the radiations from radioactive substances. Supplemented by chemical methods[c] they provide information concerning the changes which occur in the materials themselves concurrently with the emission of the radiations. Precisely similar methods are employed when attention is chiefly concentrated on the changes produced by these radiations in matter on which they fall. In such cases the amount of material actually affected is always much too small for purely chemical methods to be of any avail;[d] any information which is obtained must, therefore, in the first place be information concerning the (secondary) radiations which are emitted in the course of the

[a] Blau, *J. Physique*, 5, 61, 1934; Taylor (H. J.), *Proc. Roy. Soc.* 150, 382, 1935; Taylor and Dabholkar, *Proc. Phys. Soc.* 48, 285, 1936.

[b] Frequently ionisation chambers are employed which contain gas at several atmospheres pressure. This modification introduces many advantages for certain types of work; cf. Hoffmann, *Z. Physik*, 42, 565, 1927; Tarrant, *Proc. Roy. Soc.* 128, 345, 1930; Gray, *ibid.* 130, 524, 1931; also Florance, *Phil. Mag.* 25, 172, 1913.

[c] Russell (A. S.), *An Introduction to the Chemistry of Radioactive Substances*, 1922.

[d] See, however, p. 126.

transformations. Obviously, the experimental arrangement must be such that these radiations shall be detected in spite of the presence of an enormously more intense primary radiation which is responsible for producing them. Sometimes no special precautions are necessary to achieve this result—in the experiments of Rutherford in 1919, when evidence for artificial disintegration was first obtained, none were required—but in other instances the most elaborate care is called for in this respect. In 1919 Rutherford[a] was using the scintillation method to examine the possibility that the passage of α particles through matter resulted in the projection of other particles by the process of collision. Experimentally it was simplest first of all to investigate whether there were any such particles projected forwards, beyond the normal range of action of the α particles. This was done by placing the scintillation screen in front of the source, interposing the material under investigation in a layer thick enough completely to cut off the direct action of the α particles, and observing whether any scintillations could still be seen. Such scintillations were, in fact, found. They were attributed to the action of long range particles produced in small numbers when the primary radiation was absorbed in matter. Although the full results were not at that time altogether free from complications, the main issue was sufficiently clear: the number and range of action of these projected particles varied from one bombarded substance to another in such an irregular manner as to suggest that the effects were indeed specific to the elements bombarded; it was simplest to suppose that in rare circumstances and with certain elements[b] an atom of the element can be permanently transformed and disintegrated through impact of an α particle. From this experiment—and inspired by the challenge of the interpretation offered in respect of it—an enormous number of similar investigations have sprung: similar in principle, that is, if considerably more complicated in technical accompaniments. Use of the scintillation screen has been replaced by electrical

[a] Rutherford, *Phil. Mag.* 37, 581, 1919.
[b] Rutherford and Chadwick, *Phil. Mag.* 42, 809, 1921.

counting methods and laborious work with the expansion chamber has been rewarded by a few successful photographs showing the paths of the particles before and after a disintegration collision. Experiment, previously limited to a total of roughly an hour a day, for fear of visual fatigue in the observation of scintillations, is now frequently almost continuous and, more important still, the permissible rates of counting have been greatly extended—by a factor of a hundred or more in each direction.[a]

§ 3. *Experimental methods in mass spectroscopy.* The method of mass spectrum analysis represents a natural development of the earliest systematic investigations of phenomena associated with the passage of electricity through gases. For a long time it had been a well-known fact that a volume of gas at low pressure acts as a conductor of electricity when a sufficiently large difference of potential is applied to electrodes introduced into the containing vessel. The fundamental problem was to formulate a clear description of the mechanism of this conduction. For this purpose the observations of Plücker[b] provide the necessary starting-point, although the full implication of these observations was not generally recognised at the time. Some twenty years later Crookes[c] first consistently maintained the view that the previously observed effect of a magnetic field on certain types of discharge was to be explained in terms of the transport of negative electricity by charged particles possessing finite inertia. Such particles would move in a magnetic field in curved paths, the constants of which would depend upon the charge and mass of the particles and upon their velocity. In 1895 Perrin[d] showed that negative electricity was in fact carried by the "cathode rays", and in 1897 their corpuscular nature was finally established, amongst other ways,[e] by the

[a] Wynn-Williams, *Proc. Roy. Soc.* 132, 295, 1931; *ibid.* 136, 312, 1932; Hoffmann and Pose, *Z. Physik*, 56, 405, 1929.

[b] Plücker, *Pogg. Ann.* 103, 88, 1858; *ibid.* 107, 77, 1859.

[c] Crookes, *Phil. Trans. Roy. Soc.* 170, 135, 641. 1879.

[d] Perrin, *Comptes rendus*, 121, 1130, 1895.

[e] See Wiechert, *Sitzungsber. d. Physikal.-ökonom. Ges. zu Königsberg im Pr.* 38, 1, 1897; Kaufmann, *Wied. Ann.* 61, 544; *ibid.* 62, 596, 1897.

demonstration—previously attempted but without complete success—of the deflection produced by a transverse electric field. This was the work of J. J. Thomson[a] and he proceeded to apply the results to a determination of the velocity and specific charge[b] of the particles. He showed that these particles were the same whatever the chemical nature of the residual gas or of the material of the electrodes. In a similar manner he demonstrated their presence in a number of different circumstances, when negative electricity was being liberated from matter. Subsequent determinations of the ionic charge[c] enabled the mass of the particles to be calculated. It was found to be of the order of 1000 times smaller than the mass of the hydrogen atom. Gradually, therefore, as a result of such experiments, the cathode particle—or "negative electron"—took its place in the conceptual scheme as an accepted common constituent of the atoms of matter.

In principle Thomson's method, which involves the deflection of a beam of fast-moving particles in known electric and magnetic fields, makes possible the analysis of the particles in any beam in respect of velocity and specific charge. An early success of the method was the demonstration[d] that the β radiation from radioactive materials consists of particles themselves identical with the electrons from the discharge tube. On the other hand, about the same time, Wien[e] was able to show that high-speed particles of atomic mass were also concerned in the transport of (positive) charge in a gas at low pressure. The "canal rays" of Goldstein[f] were shown to be of this nature: their analysis by the method of Wien was pursued with great success by Thomson, chiefly during the period 1906 to 1920.[g] For present purposes the most im-

[a] Thomson, *Phil. Mag.* 44, 293, 1897.

[b] The quantity e/m—which may only loosely be referred to as the ratio of charge to mass—is sometimes spoken of as the specific charge of a particle of mass m and charge e.

[c] Townsend, *Proc. Camb. Phil. Soc.* 9, 244, 1897; *Phil. Mag.* 45, 125, 1898; Thomson, *Phil. Mag.* 46, 528, 1898; *ibid.* 48, 547, 1899.

[d] Becquerel, *Comptes rendus*, 130, 809, 1900.

[e] Wien, *Wied. Ann.* 65, 440, 1898.

[f] Goldstein, *Berl. Sitzungsb.* 39, 691, 1886.

[g] Thomson, *Rays of Positive Electricity*, 1921.

portant result of these experiments may be stated as follows: in terms of the value $(e/m)_H$, for the specific charge of the atomic hydrogen ion, as unit, the specific charge of any ion may with some accuracy be expressed as an integral proper fraction q/A, where q has most frequently the value 1 or 2 and in common experience never a value greater than 8. This is at once a demonstration of the different masses of the atoms of the elements, their very closely integral values in terms of the chosen unit,[a] and of the fact that the positive charge carried by an atomic ion is always an integral multiple of the electronic charge.[b] It results from these considerations that, apart from the slight ambiguity involved in the different values possible for q in any case, the method applied by Thomson may be employed for the analysis, in respect of mass only, of the ions constituting any well-defined beam. The information obtained with the original apparatus—and more particularly the effect observed with neon[c]—provided the incentive for the application of more powerful methods of a similar nature. Thus Aston[d] and, later, Bainbridge[e] and others constructed instruments of high (mass) dispersion and great resolving power. These instruments have been progressively improved during recent years.[f] For the lighter elements, at least, it is now possible to determine atomic (ionic) masses with an error of less than 1 part in 10,000—a most important fact, of which the full bearing will be discussed at length in chapter V. For the present, however, it will be useful to consider in a little more detail the development of the general experimental method involved.

[a] A much better representation of specific ionic charge values by an integral proper fraction, as above, is obtained if this unit is taken as sixteen times the value of the specific charge of the most abundant positive ions of atomic oxygen. On this basis the integer A, in any case, is referred to as the mass number of the ion (nucleus) in question. On the former (hydrogen) basis the specific charge of the important oxygen ion is about 1/15·88.

[b] When a similar suggestion was made in respect of electrolytic ions, in order to account for the facts of electrolysis, Maxwell (*Electricity and Magnetism*, 3rd edition, 1, 379, 1892) was amongst those who did not regard it with complete favour.

[c] Thomson, *Proc. Camb. Phil. Soc.* 17, 201, 1913; *Proc. Roy. Soc.* 89, 1, 1913.

[d] Aston, *Phil. Mag.* 38, 707, 1919.

[e] Bainbridge, *Phys. Rev.* 40, 130, 1932; *J. Franklin Inst.* 215, 509, 1933.

[f] Aston, *Mass Spectra and Isotopes*, 1933.

Suppose that a particle, of mass m and charge e, is moving with velocity v in a uniform electric field of strength X, at right angles to the direction of X. It will possess an acceleration Xe/m parallel to the direction of the field. If, now, the electric field is replaced by a magnetic field of strength H, also at right angles to the direction of motion of the particle, the instantaneous magnitude of the acceleration becomes Hev/mc—and its direction at right angles both to the field and to the direction of motion. In either case the deflection suffered by the particle in travelling a known distance in the field may be calculated. If these deflections are small compared with the length of path in the field, then, when both fields act together, over the same distance, the displacement of the particle at any point in its path will be the resultant of the corresponding displacements due to the two fields acting separately and the final deflection will similarly be the resultant of the deflections obtained separately. Most of the experimental arrangements which have been employed make use of this result regarding the simultaneous effects of two analysing fields; the essential difference between one and another is just a difference in disposition of the fields, with the resultant difference in the type of deflection pattern obtained when a beam of ions of many kinds and different velocities is submitted to analysis.

With cathode rays Thomson employed electric and magnetic fields at right angles, so arranged as to produce opposing deflections of the beam. In such a case particles having the favoured velocity $v_0 = cX/H$ pass without deflection through the fields, whilst, for velocities greater or less than v_0, the resultant deflection (parallel to the axis of X) is to one side or the other depending upon the sign of e.[a] Conversely, when (as is usually the case with a considerable fraction of the cathode rays) the particles are approximately homogeneous with respect to velocity, their common velocity may be obtained by the same method. Field strengths having been adjusted to give zero resultant deflection of the beam, the ratio X/H is a measure of v/c. In this way Thomson

[a] In this connection the arrangement may be spoken of as a velocity filter.

determined the velocity of the cathode particles; by measuring the deflection of the same beam due to a known magnetic field acting alone he was then able to calculate the value of e/m for the particles. The close similarity of this method with that of Bainbridge for high-speed positive ions will be evident at a later stage (p. 17). Meanwhile it may be useful to show —in support of the general contention of § 1 of this chapter— exactly how the atomic constants deduced by calculation in Thomson's investigation depend in the final analysis entirely upon measurements of length, mass and time on a macroscopic scale. Measurement of the linear dimensions of the apparatus and of the deflection is obviously of this kind—the lengths involved here are of the general order of magnitude of one centimetre—similarly in any determination of the velocity of light (which enters into the calculation) lengths of this order and greater have to be measured and small periods of time are calculated from observations extending over many minutes. It remains only to analyse the methods employed for the determination of magnetic and electrical field strengths to become convinced that the above statement is true in respect of them also. When the magnetic field is produced by an electric current flowing in a Helmholtz arrangement of coils, as in Thomson's experiment, its intensity may be calculated in terms of the linear constants of the coils and the strength of the current. The magnitude of the latter in electromagnetic units may be supposed measured by means of the current balance, that is in terms of the known force due to a number of milligram "weights" and a further set of geometrical constants. In electrostatic units, again, the electric field intensity X is calculated from measured lengths and the counterpoise weight of an attracted disk electrometer. Weight is the last remaining derived magnitude in either case. Here the comparison of masses is involved and a knowledge of the acceleration due to gravity—but, then, the determination of g is, perhaps, the best known example of measurement in terms of macroscopic lengths and times. At this stage the analysis of the original method may be regarded as complete.

Any other method which similarly involves the use of analysing fields disposed at right angles is obviously equivalent to that just discussed whether the deflections due to known fields are separately measured or one is balanced directly against another which has previously been determined. This is so obvious a corollary that to mention it would have been unnecessary but for the fact that the first determination of velocity and specific charge for α particles was carried out in this way. Rutherford[a] determined the deflections of the canalised α radiation from a thin layer of radioactive material due to the action of transverse electric and magnetic fields separately applied. He showed that, when the material was chemically of one kind, in general α particles of a single velocity only were emitted, although this characteristic velocity differed from one radioactive substance to another.[b] The specific charge of the α particles, however, was always the same—roughly that appropriate to an atom of helium carrying two positive charges. The true nature of radioactivity was thus beginning to be evident: it might be described as the spontaneous emission of high-speed particles from the atoms of unstable chemical elements. In disintegrations of one type (β disintegrations) particles of negative charge and sub-atomic mass were emitted, in those of another (α disintegrations) particles of light atomic mass and positive charge were expelled. The attempt to produce deflection of the γ radiation by electric and magnetic fields completely failed; this radiation was, therefore, rightly regarded as uncharged, although two different views concerning its other attributes were variously sponsored for a number of years.

The use of transverse analysing fields simultaneously applied over the same region of space and in the same direction must be attributed in the first place to Kaufmann,[c] who employed this arrangement for his investigations of the cathode rays. It was then adopted by Wien[d] for canal rays

[a] Rutherford, *Phil. Mag.* 5, 177, 1903.
[b] Rutherford, *Phil. Mag.* 10, 163, 1905; *ibid.* 11, 166; *ibid.* 12, 134, 348, 1906.
[c] Kaufmann, *Phys. Z.* 2, 602, 1901. [d] Wien, *Ann. Physik*, 8, 244, 1902.

and used by Thomson in his extensive work on the same subject, to which some reference has already been made. The deflection pattern, here, for a receiving surface perpendicular to the original direction of the ionic beam consists of portions of each of a set of parabolas having axis and vertex in common. Any one parabola of the set is the locus of the points of arrival of ions characterised by a single value of e/m; the distribution of intensity along such a parabola is determined by the distribution of initial velocity amongst the ions. Measurement of the geometrical latus rectum for any parabola enables the corresponding specific charge to be calculated, once electric and magnetic field strengths are known and the linear dimensions of the apparatus have been determined. Obviously this process may be carried out with the greatest accuracy when a considerable portion of the parabola appears developed in the pattern. Unfortunately this is not always the case; concentration of intensity in the parabola "head" is the inevitable result of the frequent discharge condition which gives rise to a beam of canal rays having velocities generally within a small range of the maximum velocity. In these cases the parabola method is the least accurate:[a] Aston's mass spectrograph, on the other hand, is most effectively employed when such discharge conditions have been obtained.

It is the virtue of this instrument that "velocity focusing" is introduced. Crossed analysing fields are applied to the moving particles over separate portions of their paths and are so disposed, together with suitable diaphragms, and adjusted as regards strength, that, for the range of velocity (energy) most represented, there is the necessary deviation of the beam but no dispersion of velocities. A photographic plate appropriately placed receives the focused rays at oblique incidence. When this is developed a line spectrum is obtained from which values of e/m may be deduced. Usually, however, absolute values of e/m are not in question—merely the num-

[a] Of recent years, however, considerable refinements have been introduced. For these see Eisenhut and Conrad, *Z. Elektrochemie*, 36, 654, 1930; Conrad, *Phys. Z.* 31, 888, 1930; Lukanow and Schütze, *Z. Physik*, 82, 610, 1933, also Zeeman and de Gier, *Proc. K. Akad. Amsterdam*, 37, 127, 1934.

ber q/A (p. 12) which relates the value of e/m for any particular ion to the specific charge of some standard ion. Concerning the various methods by which this comparison is effected reference must be made to the original papers: suffice to say here that although it is obviously important that these measurements be carried out with the utmost refinement of method and the highest accuracy, yet they do not constitute absolute determinations of any atomic quantity. If the need for an accurate absolute determination of specific charge should arise, the method of Bainbridge is more easily adaptable to this end.

Bainbridge's method—which like Aston's is normally used for comparative measurements only—may be regarded as an improvement of an earlier method due to Dempster.[a] In each case magnetic analysis alone is applied to a beam of positive ions, heterogeneous with respect to mass and charge, which has previously been rendered effectively homogeneous with respect to energy or velocity. Constituent beams being bent into circular arcs in the field, observed curvatures then serve to determine the quantities $e/m^{\frac{1}{2}}$ or e/m, as the case may be. Dempster attempted to produce ions of a single energy only by acceleration in a known electric field, Bainbridge adopted the velocity filter of Thomson (p. 13) to transmit ions of very nearly a single velocity. The latter condition may be more nearly realised in practice. For the magnetic analysis both Dempster and Bainbridge employed the semi-circular focusing method of Danysz[b] as modified by Rutherford and Robinson.[c] This[d] depends upon a geometrical property of equal circles—and, of course, upon the fact that the free path of a charged particle in a magnetic field is a circular helix with its axis along the direction of the field. The geometrical property may be stated as follows. If two lines PA, PB, including a small angle, be drawn through a point P and, in the plane so defined, any number of equal circles be drawn

[a] Dempster, *Phys. Rev.* 11, 316, 1918. [b] Danysz, *Le radium*, 10, 4, 1913.

[c] Rutherford and Robinson, *Phil. Mag.* 26, 717, 1913.

[d] It appears that the same method had previously been employed by Classen (*Phys. Z.* 9, 762, 1908) without its focusing action having been explicitly pointed out.

through P, such that tangents through P to these circles all lie within the angle APB, then these circles again converge in a very small area traversed by the external bisector of APB. Good focusing is therefore possible when a finite bundle of rays from a line source is passed through a moderately wide slit of which the length is parallel to the length of the source and also to the axis of the analysing field. Particles having a small component of initial velocity parallel to the field, as well as those for which this component is effectively zero, add to the general intensity of the "spectrum line".

The semi-circular focusing method, as above indicated, was developed chiefly by Rutherford and Robinson. It was employed by them, and by Ellis, Meitner and many others following them, for the investigation of the distribution of velocity amongst the β particles from various radioactive sources. Strictly speaking, it is not a method of mass spectroscopy, when so used; it gives a magnetic velocity spectrum[a] for particles of one kind for which e/m is known. It is mentioned here, however, for lack of another place and because of the natural connection with Bainbridge's method. Until recently it was employed for electrons only; the magnetic velocity spectrum for heavy particles had not similarly been obtained by the focusing method. For α particles direct deviation methods were used, following Rutherford[b] and Rutherford and Robinson,[c] and in these the width of the spectrum line must always be greater than the width of the slit. Latterly, however, the outstanding difficulty—the maintenance of sufficiently intense magnetic fields over sufficiently extended regions of space, in vacuo—has been overcome, and both in Paris[d] and in Cambridge[e] very accurate determinations of α particle velocities have been made by this means.

[a] The term "magnetic spectrum" was first used in a similar connotation by Birkeland, in reference to cathode rays (*Comptes rendus*, 123, 492, 1896).

[b] Rutherford, *loc. cit.* p. 15.

[c] Rutherford and Robinson, *Phil. Mag.* 28, 552, 1914.

[d] Rosenblum, *Comptes rendus*, 188, 1401, 1929.

[e] Rutherford, Wynn-Williams, Lewis and Bowden, *Proc. Roy. Soc.* 139, 617, 1933.

Analysis of beams of fast moving charged particles by applied fields may thus be made to give specific charge (or mass) and velocity of the particles;[a] it may also be arranged to provide information concerning atomic magnetic moments, if uncharged particles are available and magnetic fields of sufficient inhomogeneity may be employed. Thus, if an electrically neutral particle of resultant magnetic moment μ be situated in a non-uniform field of strength H at a given point, when the axes of μ and H are inclined at an angle θ, the particle is acted upon by a force $\mu \cos \theta \left(\frac{\partial H}{\partial x}\right)$ in the direction of measurement of x. This result was first utilised by Gerlach and Stern[b] in the design of an apparatus for the determination of atomic magnetic moments by the deflection of atom beams. In practice its application has been limited only because of the necessity for sufficiently intense beams of uncharged particles, which must be produced in most cases by evaporation from suitable sources. The relevance of the method here consists in the fact that by further refinement of the original arrangement it is possible to determine also nuclear spin moments in favourable cases (see chap. VI). Stern[c] and Rabi[d] have been responsible for the methodological refinements in question.

§ 4. *Relevant experimental methods in optical spectroscopy.* The optical spectroscope is essentially an instrument for analysing visible or near-visible light in respect of those characteristics which determine perceived colour. Quantitatively, it provides the means of measuring wave-lengths (λ) of the quasi-homogeneous components of the light incident upon it—most directly in terms of the fundamental interval of a linear

[a] A new method of magnetic velocity analysis for electrons has recently been described by Klemperer (*Phil. Mag.* 20, 545, 1935) in which the focusing property of localised magnetic fields for divergent, axially incident, beams of moving charged particles is employed. In this instrument the field coils may be referred to as "electron lenses" and a complete scheme of analogy between it and the optical spectroscope may similarly be worked out. (See also Hughes. *Phil. Mag.* 19, 129, 1935.)

[b] Gerlach and Stern, *Z. Physik*, 8, 110, 1922; *ibid.* 9, 349, 353, 1922.

[c] Frisch and Stern, *Z. Physik*, 85, 4, 1933.

[d] Breit and Rabi, *Phys. Rev.* 38, 2082, 1931.

grating. From these wave-lengths, oscillation frequencies (ν) may be calculated using the relation $\lambda\nu = C$, C being the velocity of propagation of light of the frequency in question through the medium to which λ refers. Whilst, in any case, λ and C depend upon the medium, frequencies are characteristic of the emitting system only.[a] In the last analysis the atoms and molecules of matter must be regarded as emitters, so that, when a source of light is such that interaction between neighbouring atoms is negligible (as in a volume of gas at low pressure) the frequencies of the emitted light appear as characteristic oscillation frequencies of single atoms. This conclusion assumes importance from the general postulate of Planck,[b] that emission and absorption of energy in the form of radiation of frequency ν takes place in amounts $h\nu$, but not in smaller amounts—the quantity h being a universal constant ($6{\cdot}59 \times 10^{-27}$ erg sec.). It appears that energy differences characteristic of atomic and molecular systems may thus be calculated from the known frequencies of the light which, under suitable excitation, they emit. For purposes of "explanation" it becomes necessary to endow the conceptual entity atom with just such a structure as will admit of its existence in excited states of excess energy and finally provide a basis for the numerical calculation of characteristic energy differences and intensity relations descriptive of the process of emission of light. From observations of the effect of a magnetic field on this emission,[c] Lorentz[d] concluded that negatively charged particles in the atom were primarily involved in the emission process. On the basis of his interpretation the specific charge of the particles could be calculated: within tolerable limits it proved to be the same as the specific charge of the negative electron. For further progress, however—and particularly for a workable suggestion regarding the disposition of positive charge in the atom—theoretical physics was destined to wait

[a] In this discussion it is supposed that there is no relative motion amongst the material systems involved.

[b] Planck, *Verh. d. D. Phys. Ges.* 2, 237, 1900; *Ann. Physik*, 4, 553, 1901.

[c] Zeeman, *Phil. Mag.* 43, 226, 1897.

[d] Lorentz, *Wied. Ann.* 63, 278, 1897.

until the nuclear hypothesis of Rutherford (1911) became the basis of Bohr's[a] spectral theory in 1913. From that date, if by means of a rather uneven advance, theory has been able at least to keep pace with spectroscopic experiment. In broad outline, results in atomic spectra are explained in terms of the emission and absorption of energy by a system of Z electrons having certain possible configurations about a massive ("fixed") centre of electrostatic force, acting as such in virtue of its compensating positive charge Ze. The mass of the electron, but not the very much greater mass of the nucleus—nor any other characteristic than its charge—enters into these approximate calculations. In the finer analysis, however, other nuclear constants enter also. Thus, for light atoms, it is no longer sufficiently accurate to regard the nuclear mass as effectively infinite, the error in characteristic energy values introduced by this assumption being large enough to be detected, and for all atoms, in theory, it is to be expected that the occurrence of nuclear rotation might similarly be detected through its effects. This latter is an important conclusion, and much experimental work under conditions of extreme optical resolving power—making possible the detection of very small differences in frequency or characteristic energy—has produced a considerable body of data which are consistently explained in terms of nuclear spin moments of varying amount. These are obviously important constants for the purpose of any theory of the nucleus. For that reason the methods in optical spectroscopy most relevant to the present subject are those of highest resolution: they involve the use of interference spectroscopes[b] in one form or another.

Certain features of molecular spectra, on the other hand, are relevant here which are not matters of extreme detail. Since, for these spectra, the moments of inertia of the molecule are determining constants, changes in nuclear mass are more easily noticeable, consequently the "isotope effect" may become a prominent feature of the spectrum.[c] Again, how-

[a] Bohr, *Phil. Mag.* 26, 1, 476, 857, 1913.
[b] Williams (W. E.), *Applications of Interferometry*, 1930.
[c] Imes, *Astrophys. J.* 50, 251, 1919; Loomis, *ibid.* 52, 248, 1920.

ever, the finest detail provides the most important information, and, as before, this information concerns the rotational characteristics of the nuclei concerned. One difference only has to be noted. This information is obtained from molecular band spectrum photographs in terms of the measured intensities of neighbouring lines, not from their separation on the scale of wave-length or frequency. The travelling microscope or comparator, which for most purposes was previously sufficient in itself, has thus to be provided with a microphotometer attachment, for full utility.

Methods of X-ray spectroscopy, whilst not falling strictly within the compass of this section, may also be briefly mentioned here, if only for their historical importance (see p. 34). They are employed in the measurement of wave-length in the case of the radiations which are emitted from material targets of all kinds under bombardment by cathode rays of high energy.[a] Most directly this measurement is made in terms of the fundamental interval characteristic of the regular spacing of atoms or ions in a crystal lattice.[b] At an early stage it was realised that the X-rays from a given target are in part characteristic of the individual atoms present;[c] it remained for Moseley[d] to show by the new methods how extremely simple the spectrum of this characteristic radiation is and to exhibit its important regularities. An explanation of this simplicity and an interpretation of the regularities in terms of nuclear charge was one of the early successes of Bohr's theory. Nowadays, however, the analysis of X-ray spectra is almost without direct interest for the nuclear physicist—and the application of essentially similar (crystal) methods to the determination of γ-ray wave-lengths has long been superseded by the use of more exact indirect methods of attack. It is interesting to note, however, that originally these crystal methods provided the most straight-

[a] Röntgen, *Wied. Ann.* 64, 1, 1898.

[b] Bragg (W. L.), *Proc. Camb. Phil. Soc.* 17, 43, 1912; Bragg (W. H.) and Bragg (W. L.), *Proc. Roy. Soc.* 88, 428, 1913.

[c] Kaye, *Phil. Trans. Roy. Soc.* 209, 123, 1909; cf. Barkla, *Phil. Mag.* 11, 812, 1906.

[d] Moseley, *Phil. Mag.* 26, 1024, 1913; *ibid.* 27, 703, 1914.

forward evidence for the identity in nature between γ rays and X-rays[a] (and visible light), which virtually decided a long controversy concerning the nature of the third component of the radioactive radiations.

§ 5. *Experiments with artificially accelerated particles.* The result which has just been quoted (§ 4) may be said to complete the experimental proof that the radiations from radioactive substances are identical in kind with similar radiations which may be produced in the discharge tube: the α particles with doubly charged helium positive ray particles, the β particles with the cathode-ray particles and the γ radiation with the X-radiation produced by impact of the cathode rays on a suitable target. The difference between the natural radiations and those "artificially" produced in this way is one of energy and intensity only. Particle energies and radiation frequencies are in general much greater when the radiation is of radioactive origin than when it is not; intensities,[b] on the other hand, are usually incomparably smaller. On the general assumption that lack of individual particle energy may sometimes and for some purposes be partially counterbalanced by greatly increased numbers of projectiles, the method of artificial acceleration of ions was first developed with a view to the investigation of atomic disintegration in the years 1929–32.[c] Amongst other things this method provided the opportunity of employing as projectiles positive particles other than α particles—and amongst such particles, most obviously, the proton, the simplest positive particle, the atomic hydrogen ion (p. 12). The success of this move may be gauged by reference to chap. XII which is entirely concerned with details of the results so obtained.

Here, two considerations of a more general character are in order. There is the consideration of the means employed for accelerating the particles, on the one hand, and, on the other,

[a] Rutherford and Andrade, *Phil. Mag.* 27, 854, 1914; *ibid.* 28, 263, 1914.

[b] Measured by the flux of particles or of radiation quanta.

[c] Breit, Tuve and Dahl, *Phys. Rev.* 35, 51, 1930; Cockcroft and Walton, *Proc. Roy. Soc.* 129, 477, 1930; *ibid.* 136, 619, 1932; Lawrence and Edlefsen, *Science*, 72, 376, 1930; Lawrence and Livingston, *Phys. Rev.* 37, 1707, 1931; Brasch and Lange, *Z. Physik*, 70, 10, 1931; Van de Graaff, *Phys. Rev.* 38, 1919, 1931.

of the methods used for detecting the products of disintegration. The second of these questions may be dealt with at once: the methods employed are essentially those previously developed in connection with the investigation of disintegrations produced by bombardment with α particles (p. 9). Thus the scintillation screen, the ionisation chamber, various forms of particle counter with appropriate amplifying systems, and the expansion chamber, have all been used. Only slight modifications in standard practice have been necessary on account of the presence near the detector of electrical machines operating at high potentials, and because of the very great intensity of primary beam which frequently is required in order to produce the effects which are of importance: the yield of disintegrations per primary particle at these speeds is often very small indeed. More particularly, in such cases, the production of general X-radiation in the positive ray tube may assume serious proportions, but, fortunately, this radiation is not of a very penetrating character.

Concerning the methods which have been used for accelerating the particles, many of the details are matters of electrotechnics simply, and will not be entered into here, but a few general divisions may be recognised. Most of the methods consist in introducing an intense beam of particles from a subsidiary discharge[a] into a long accelerating tube[b] at the end of which the target is situated—either inside the main tube or separated from it by a thin window. The difference of potential applied to the tube may be steady or oscillating and the resulting high speed particles may or may not be subject to rough magnetic (mass-velocity) analysis before being directed on to the target. If this analysis is not carried out the final beam most frequently will be composed of particles of different kinds—atomic and molecular ions together with some particles which are uncharged. For the best results acceleration by a steady potential difference, followed by

[a] Cockcroft and Walton, *loc. cit.*; Oliphant and Rutherford, *Proc. Roy. Soc.* 141, 259, 1933; Tuve, Dahl and van Atta, *Phys. Rev.* 46, 1027, 1934.

[b] Coolidge, *J. Franklin Inst.* 202, 693, 1926; Tuve, Breit and Hafstad, *Phys. Rev.* 35, 66, 1930; Gedye and Allibone, *Proc. Roy. Soc.* 130, 346, 1931.

magnetic analysis of the beam, is obviously to be recommended, all other considerations being equal. Some methods, however, produce high speed particles by successive acceleration in such a way that large potential differences are not necessary. For many purposes, evidently, these methods have great advantages.

Two methods of using steady potentials have been successfully developed, by Cockcroft and Walton and by Van de Graaff, respectively. In the first of these the current from a high-tension transformer is used to supply a system of condensers through an arrangement of rectifying valves in such a way that the main condenser is finally charged to a difference of potential nearly four[a] times as great as may be obtained directly from the transformer. During continuous working current may be taken from this condenser at approximately constant voltage, if the load is not too great and the constants of the circuit have been suitably chosen. The high voltage generator of Van de Graaff, on the other hand, is based on principles of electrostatics. Point discharges at a fairly low potential are employed to produce surface electrifications of opposite sign on two "endless" belts of insulating material running between "earth" and the centres of two hollow metal spheres suitably insulated. By means of collecting brushes inside these spheres the charges conveyed by the belts are transferred to the electrodes which thus acquire a great difference of potential. The current which may be taken from them at this electrical pressure depends upon the belt speeds and their width, a maximum density of surface electrification of the belts (determined by the dévelopment of brush discharge from the edges) being easily attained with the apparatus available. The method is essentially one in which small currents must be offset by the possibility of reaching very high voltages. It appears that for this possibility to be realised it is only necessary to increase correspondingly the size of the apparatus employed.[b]

[a] The voltage multiplying factor may be given any even value if the number of stages in the electrical circuit is chosen correspondingly.

[b] Van de Graaff, Compton (K. T.) and van Atta, *Phys. Rev.* 43, 149, 1933.

Of the methods which involve the use of varying potentials probably the greatest amount of work has so far been done by the method of Lauritsen. In this case a million-volt transformer was available[a] and it was merely necessary to modify its previous mode of application in the production of X-rays (for deep therapy)[b] to the purpose in hand.[c] Perhaps because of these antecedents, in this particular case most of the earlier investigations dealt with the γ rays rather than with the particles emitted in the course of artificially produced nuclear transformations. Extremely high potentials may also be obtained by use of the Tesla coil or oscillation transformer: this is the method of Breit, Tuve and Dahl. In its usual form, however, it is a method of low average intensity and no particular advantage—the high frequency oscillations are excited for a very small fraction only of the total available time. Sloan[d] has attempted to overcome this disadvantage by exciting the primary of the transformer by means of maintained oscillations, but for this a larger input of power is required. The originators of the method, Breit, Tuve and Dahl, have abandoned it for a modification of the electrostatic generator of Van de Graaff.[e] There are some purposes, however, to which high intensity ion beams which are available only for very short times are entirely suited: investigations by means of the expansion chamber[f] are in this class. For them a more completely discontinuous type of excitation even than that of the Tesla coil may sometimes be advantageous. Such an arrangement is provided by the impulse generator described by Brasch and Lange. Up to the present time it has not, in fact, been used to this end, but very compact impulse sets constructed for quite other purposes[g] have

[a] Sorensen, *Amer. Inst. E. Eng.* 44, 373, 1925.

[b] Lauritsen and Bennett, *Phys. Rev.* 32, 850, 1928.

[c] Crane, Lauritsen and Soltan, *Phys. Rev.* 45, 507, 1934.

[d] Sloan, *Phys. Rev.* 47, 62, 1935.

[e] In this modification (Tuve, Hafstad and Dahl, *Phys. Rev.* 45, 768, 1934; *ibid.* 48, 315, 1935) the two spherical electrodes are made concentric and of considerably different radii, so that the discharge tube may be inserted between them. A great saving of space is thereby effected.

[f] With an automatic expansion chamber of normal design about $\frac{1}{20}$ to $\frac{1}{50}$ sec., every 15 to 30 secs., is all the time which is effective.

[g] Allibone, Edwards and McKenzie, *Nature*, 131, 129, 1933.

demonstrated the possibilities of the method. It consists essentially in charging up a battery of condensers in parallel to a low voltage and discharging them in series. In this way the voltage multiplication of the arrangement is limited only by the number of condensers constituting the battery.

Various methods of production of high speed particles without the use of correspondingly great differences of potential have been tried from time to time. The earliest attempts in this direction, by Wideroe[a] and Walton,[b] were concerned with electrons: they were not very successful. A much greater measure of success, however, has since been achieved with positive ions. This is natural enough, for the following reason. The methods in question depend essentially upon the reversal in sign of an accelerating field during the time taken by the particle to travel in a "field free space" from one accelerating gap to another. Now, for a given apparatus, this time is longer the heavier the particle (for the same energy), and in a high-power oscillating circuit the longer the period the more easily are the oscillations maintained. Mercury ions may be taken as an extreme instance of this. In this case, over a total path of a few metres, Sloan[c] has been able to arrange 36 tubular elements in line, which when connected alternately to the output terminals of a 79,000-volt oscillator of suitable frequency produce a resulting ion beam of nearly 3.10^6 electron volts energy. This arrangement of successive elements in line, however, becomes altogether too long when the lightest positive ions are in question, and it is with them that the most significant experiments in nuclear physics are likely to be concerned for some time to come. For that reason, amongst the methods which do not require large differences of potential, that of Lawrence is receiving the greatest attention. Here use is made of the fact that, in a plane at right angles to the direction of a uniform magnetic field, charged particles move in circles of radii (ρ) proportional to the appropriate values of

[a] Wideroe, *Archiv f. Electrotechnik*, 21, 287, 1928.
[b] Walton, *Proc. Camb. Phil. Soc.* 25, 469, 1929.
[c] Sloan and Lawrence, *Phys. Rev.* 38, 2021, 1931; Sloan and Coates, *ibid.* 46, 539, 1934.

mv/e. For particles of one kind, therefore, $2\pi\rho/v$ is independent of v, or the time of description of the circular path is independent of the velocity. Because of this fact a single pair of electrodes may be used to replace the linear array of the direct method. They may be regarded as formed by a diametral saw-cut through a shallow cylindrical box with closed ends. The ion source is placed within and near the centre of this divided box, the magnetic field is applied along the axis and the output leads of the oscillator are connected to its two halves. When the period of the oscillations is made equal to the time taken by the particles to complete one circuit round the direction of the field, any one particle passes through the gap between the plates twice in a complete period, receiving on each occasion equal increments of energy. Each particle, therefore, gradually spirals out towards the periphery of the box with regularly increasing energy. Finally, by means of a subsidiary electrostatic field, it is withdrawn through a hole in the cylindrical wall and directed on to whatever target is under investigation. Beams of deuterons of about $5 . 10^6$ electron volts energy have thus been obtained by use of an oscillator producing 20,000 volts on output.[a] Another method of acceleration which does not involve the use either of high potentials or of a set of accelerating gaps and intermediate shielded regions has been described by Beams and Snoddy.[b] It consists essentially of a suitably loaded transmission line along which an electrical impulse may be made to travel so as to keep pace with a cloud of ions which thus acquire considerable energy through being continuously acted on by the impulsive field. As yet, however, this method has not been employed in any experimental attack on the problems dealt with in this book.

[a] Lawrence and Livingston, *Phys. Rev.* 45, 608, 1934.

[b] Beams and Snoddy, *Phys. Rev.* 44, 784, 1933; Snoddy, Beams, Ham and Trotter, *Nature*, 138, 167, 1936.

CHAPTER II

INTERPRETATION WITHIN THE SCHEME OF MACROSCOPIC PHYSICS

§ 6. *The "classical" background.* For many years fairly successful attempts were made to interpret the results of experiments such as those already described in terms of ideas which are frequently referred to as constituting the scheme of "classical" physics. It is necessary to state here precisely what these ideas are and to indicate the measure of success attained in the interpretations.

Macroscopic phenomena may be regarded as determined by the interaction of solid bodies for the one part and by the behaviour of fluid substances for the other: from these two aspects of the phenomena the ideas of material particle and of wave motion in a continuous medium naturally derive. By further abstraction the latter idea may be developed so as to apply to other than directly perceived mechanical undulations of gross matter, even if the medium necessary to sustain the wave motion be thought of as possessing no other attributes than those of extension in space and time. Thus Maxwell's electromagnetic theory of light has a place in the scheme.

Quantitative experiments led to the enunciation of laws of conservation: measurable quantities—mass, momentum, energy, electric charge—were defined in respect of particles[a] and it was recognised from common experience that the total mass, momentum, energy or charge (this total being obtained by vector or scalar addition as the case may be) in any "isolated" system is unchanged by physical processes taking place wholly within the system. Common experience, however, does not extend to those cases in which material systems are in relative motion with very large velocities.

[a] In the development of the idea of wave motion for the purposes of Maxwell's theory it is still assumed that energy may be transferred from one material system to another by means of the wave motion, in spite of the absence of a material medium.

Here crucial experiments by physicists and astronomers have resulted in a modification of the general point of view—fundamental in many respects, but, in practice, more often than not one which is concerned with entirely trivial differences. Two of the conservation laws, however, were amalgamated, since it was recognised that mass and energy are alternative aspects of a more general mass-energy attribute of matter.

This, then, is the classical background against which the earlier experiments in radioactivity and allied branches of physics were always regarded. It has been invoked in a number of cases already throughout chap. I. Three names may be given as suggesting more tersely than is otherwise possible its ideological content: the names are those of Newton, Maxwell and Einstein.

§ 7. *The nuclear hypothesis.* It was early discovered that cathode and β rays in their passage through matter are rapidly diffused, that is that individual particles experience sudden changes in direction of motion in the process. Rutherford[a] first showed that the same is true, though to a less noticeable extent, in the case of α particles. It was natural to ascribe these effects to the action of electrostatic forces, operative between the particles and the constituent parts of the atoms of matter through which they pass, similarly it was recognised that the great disparity in mass between α particle and electron was somehow involved in any quantitative explanation of the effects.[b] When all experiments were carried out with large numbers of particles, however, and only statistical results were available, it was obviously difficult to draw valid conclusions regarding the individual changes in direction of single particles, but it

[a] Rutherford, *Phil. Mag.* 11, 166, 1906.

[b] The (small) deflections suffered by two equally charged particles in passing through the same electric field are inversely as their kinetic energies. Neither the charges nor the kinetic energies of α particles and β particles are sufficiently different, however, for this simple consideration to explain completely the results concerning scattering. The masses of the particles enter particularly because the seat of the deflecting fields is also the seat of mass comparable with that of the particles (cf. p. 33)

appeared from the early experiments of Geiger[a] that in the case of α particles these changes were always small: the resultant effect through a finite thickness of matter—usually small in itself—could be regarded as compounded of a large number of smaller deflections completely independent one of another. The experiments were carried farther by Geiger and Marsden.[b] Then it was found for the first time that large resultant deflections—of 90° and more—might occasionally be observed. It was evident that the same explanation could not apply to the distribution of resultant deflections with angle over the whole range: the idea of completely independent small deflections was quite inadequate to explain these large resultants. Infrequent as they appeared to be they would have been incomparably less frequent had they been due to random favourable combinations of slight changes in direction. It was necessary to conclude that deflection through a large angle might result from a single close approach of an α particle and an atom. Now the mechanical conservation laws require that in such a "collision" the "struck" particle shall have a mass greater than that of the incident particle if the latter is to be deflected through more than 90° from its original direction of motion. Single large angle deflections of α particles, therefore, appeared to call for explanation in terms of the existence of massive centres of electrostatic force as elements in the constitution of atoms. Now at that time (1910) the accepted atom model was quite vague as to the disposition of mass and of positive charge in the atom—vague in the sense that the suggestions which were made were determined more by considerations of mathematical simplicity than by the demands of ascertained fact. It was generally concluded that a very small fraction only of the total mass could be carried by negative electrons, say $\frac{1}{2000}$ of the atomic mass, but whether the main portion was similarly associated with electric charge or not remained a matter for speculation. Thomson[c] argued, on the basis of a particular

[a] Geiger, *Proc. Roy. Soc.* 81, 174, 1908; *ibid.* 83, 492, 1910.

[b] Geiger and Marsden, *Proc. Roy. Soc.* 82, 495, 1909.

[c] Thomson (J. J.), *Phil. Mag.* 11, 769, 1906.

theory of optical dispersion, that it was in some way associated with the positive charge in the atom. For mathematical convenience this positive charge was normally regarded as occupying the whole volume assigned to the atom. As above indicated, observations on the scattering of α particles provide the experimental material by which to test any hypothesis as to the spatial distribution of that charge with which the mass of the atom is assumed to be associated. Simple numerical considerations showed that the then accepted hypothesis was altogether untenable: a much more concentrated distribution of mass and charge had obviously to be assumed. This is the nuclear hypothesis of Rutherford.[a]

Let it be supposed that experiments show that an α particle of charge $2e$ and kinetic energy eV[b] (mass M, velocity v), passing near (or through) a heavy atom, may acquire transverse momentum of amount not small compared with Mv, being thus deflected through a considerable angle. In agreement with the above conclusions this effect is ascribed to electrostatic interaction with a massive centre of force carrying a charge Ze, the sign of which remains to be determined. When the separation of particle and centre is r the electrostatic force is of the order of $2Ze^2/r^2$, depending somewhat upon the spatial distribution of the charges. If now r represents the closest distance of approach of the particles, and if the velocity of the α particle at this stage is still of the order of v, this force may be regarded as acting for a time not long compared with r/v. The transfer of momentum is then roughly $2Ze^2/r^2 . r/v$, and we have

$$2Ze^2/rv \sim Mv$$

or

$$Ze/r \sim V.$$

Now $e = 4{\cdot}8 . 10^{-10}$ e.s.u., approximately, and as a typical value for V we may write $cV/10^8 = 4{\cdot}8 . 10^6$ volts, with $c = 3 . 10^{10}$, to obtain $V = 1{\cdot}6 . 10^4$ e.s.u. Then

$$r \sim 3Z . 10^{-14} \text{ cm.} \qquad \text{......(1).}$$

[a] Rutherford, *Phil. Mag.* 21, 669, 1911.

[b] In this case the energy of the particle is commonly given as $cV/10^8$ electron volts, V being expressed in absolute electrostatic units.

Naturally, r increases linearly with Z, but to give to r a value of the order of atomic diameters (small multiples of 10^{-8} cm.) would require that Z should be of the order of one million. This was recognised as altogether improbable: the only detailed evidence for charge within the atom suggested a total negative charge, due to electrons, of the order of Ae, A being the atomic mass number with reference to hydrogen (cf. footnote, p. 12). To make the simplest hypothesis, therefore, was to identify the charge Ze effective for α particle scattering with the positive charge necessary to neutralise the total electronic charge in the atom. Then the result (1) shows that this charge, and a large fraction of the mass of a moderately "heavy" atom, must be regarded as concentrated within a sphere of radius 10^{-12} cm. The same reasoning leads to a similar conclusion regarding the charge and mass of the α particle. Moreover, since the latter charge was already known to be positive, the case of the α particle appeared naturally as a particular example of a general rule; the α particle is the free nucleus of the helium atom. In the absence of any evidence to the contrary, the whole mass of the atom was assumed to reside in the nucleus—and compensating electrons.

Very quickly the nuclear hypothesis was put to strict numerical test by Geiger.[a] On the basis of this hypothesis Rutherford had calculated the probability, q, that an α particle of kinetic energy W shall be scattered through an angle greater than ϕ in passing normally through a small thickness t of heavy matter containing n atoms per unit volume. He found

$$q = \pi n t \frac{Z^2 e^4}{W^2} \cot^2 \frac{\phi}{2} \qquad \text{......(2).}$$

Here Ze is the nuclear charge. For α particles of a definite energy and with gold as scattering material, Geiger verified the predicted variation of amount of scattering with angle over a considerable range. Subsequently Geiger and Marsden[b] extended the angular range investigated and in addition showed that in a given angular domain the amount of scatter-

[a] Geiger, *Proc. Manch. Lit. Phil. Soc.* 55, 20, 1911.

[b] Geiger and Marsden, *Phil. Mag.* 25, 604, 1913.

ing by thin screens of aluminium, copper, silver and gold was in each case proportional to the thickness of the screen. In a similar way the scattering by a thin gold foil was shown to vary as $1/v^4$, v, the velocity of the α particles, being determined from the known relation between velocity and range. Finally, the variation of scattering, under otherwise identical conditions, with the chemical nature of the scattering substance, and its absolute amount determined for one element (gold), confirmed roughly the dependence on Z^2, if for all substances examined $Z \sim \frac{A}{2}$. These experiments were all carried out by the method of scintillations: they proved in a most convincing manner the essential validity of the nuclear hypothesis and established in the process three postulates upon which the precise form of equation (2) depends. Thus the observed effects of thin films were shown (for all but very small angles of scattering) to be due to "single scattering"[a] of the particles, with considerable accuracy α particle and nucleus could evidently be regarded as point charges acting upon one another according to the inverse square law of force, uninfluenced by the surrounding electrons, and thirdly, for atoms heavier than that of aluminium, the nucleus might conveniently be regarded as a fixed centre of force. At this stage a very simple and attractive suggestion was made by van den Broek.[b] Pointing out that the ordinal number of a chemical element in the periodic table was in most cases very little different from $A/2$, he suggested that this ordinal "atomic number" gave precisely the number of electrons in the atom and thus fixed the value of its nuclear charge. This point of view was adopted by Bohr,[c] then in the process of developing his dynamical atom model on the basis of quantum principles.[d] Some months later Moseley's results (p. 22)

[a] By this it is implied that no particle is more than once scattered through any considerable angle in traversing the scattering material. Equation (2) is clearly invalid for very small values of ϕ: no expression which represents completely the results of experiment can allow $q>1$ for $\phi=0$.

[b] van den Broek, *Phys. Z.* 14, 32, 1913. [c] Bohr, *Phil. Mag.* 26, 476, 1913.

[d] This development is here regarded as still belonging essentially to classical physics: the principles of macroscopic dynamics are retained, even though the corresponding principles of electrodynamics are abandoned.

concerning the characteristic X-ray spectra of the elements provided a complete verification not only of the essential correctness of the suggestion of van den Broek, but also, it appeared, of the central assumptions of Bohr's theory.[a] The nuclear charge number obviously increased by unity in passing from one element to the next in the periodic table; on very plausible assumptions, ascribing a slight disturbing effect to the innermost electrons in the atom,[b] it could be identified with the atomic number of van den Broek. By further refinement of the conditions in the α particle scattering experiment Chadwick[c] was able to establish this result unambiguously for the elements copper, silver and platinum.

The scattering relation (2), obtained, as above mentioned, on the supposition of a massive ("fixed") scattering centre, was later generalised by Darwin[d] for the case in which the masses of α particle and nucleus, M and m, respectively, are of the same order of magnitude. Then, if $M \leqslant m$, (2) is replaced by the result

$$q' = \pi n t \frac{Z^2 e^4}{W^2} \left(\cot \phi + \sqrt{\operatorname{cosec}^2 \phi - \frac{M^2}{m^2}} \right)^2 \quad \text{......(3).}$$

Obviously $q' < q$ for all values of ϕ, thus, in so far as single scattering is concerned, the effect of the motion of the struck nucleus is to tend to concentrate the scattered α particles in the forward direction. When $M > m$ this tendency is more marked still and no α particle may be deflected through an angle greater than Φ, where

$$\sin \Phi = m/M \quad \text{......(4).}$$

In this case it is both simpler analytically—and also more directly applicable to the experiments which are to be discussed—to give the calculated distribution for the projected nuclei rather than that of the scattered α particles. If θ is the angle of projection of the nucleus (with reference to the

[a] On this point see correspondence in *Nature*, 1914—Lindemann, Nicholson and others, also "Discussion on the constitution of the atom", *Proc. Roy. Soc.* 90, 1914—appendix.

[b] For the suggestion that the characteristic X-rays are emitted by the most tightly bound electrons in the atom see Thomson, *Phil. Mag.* 23, 449, 1912.

[c] Chadwick, *Phil. Mag.* 40, 734, 1920.

[d] Darwin, *Phil. Mag.* 27, 499, 1914.

direction of initial motion of the α particle), the probability, Q, that such a particle shall set in motion a nucleus for which the angle of projection is less than θ may be shown to be

$$Q = \pi n t \frac{Z^2 e^4}{W^2} \left(\frac{M+m}{m} \right)^2 \tan^2 \theta \qquad \ldots\ldots(5),$$

where Z, n and t, as before, refer to the material and thickness of the film of matter normally traversed.

Experiments by Rutherford[a] first showed that equation (5) is quite inadequate to explain the effects observed when a beam of α particles of known energy is allowed to pass through gaseous hydrogen. Various explanations of this result might possibly be offered. However, from a consideration of expansion chamber photographs[b] it is now known that one of these explanations is certainly untenable. The discrepancies so discovered are not to be attributed to any serious failure of the conservation laws, which appear to be valid.[c] So much was tacitly assumed at the time (1919): from the first the new results were taken as evidence for departures from the inverse square law of force between the particles. During the next decade the results of many experiments on "anomalous scattering" were discussed from this point of view.[d] For all elements investigated, from hydrogen to aluminium, some evidence of the effect was found: that this evidence became less conspicuous, the heavier the element, is to be ascribed almost certainly to the limitation as regards α particle velocity imposed by the radioactive sources which are available. With faster α particles it is to be supposed that anomalous scattering would be observed with still heavier elements.

In proposing any consistent explanation of all the observations it is evidently desirable to begin with the simplest case. That the law of force between nuclei in practice is not the "classical" law for point charges may be due either to the

[a] Rutherford, *Phil. Mag.* 37, 537, 1919.

[b] Blackett, *Proc. Roy. Soc.* 103, 62, 1923; Blackett and Hudson, *ibid.* 117, 124, 1927.

[c] Equation (4) depends only on these laws and is known to represent actual fact.

[d] Rutherford, Chadwick and Ellis, *Radiations from Radioactive Substances*, 1930, chap. IX.

absolute invalidity of that law in the domain under consideration or to the fact that the charges involved may not be regarded as point charges. Any explanation which follows the second course—certainly the first to be explored—must obviously be based on acceptable assumptions regarding the general structure of nuclei. Now the structure of the α particle alone is in question when anomalous scattering in helium is to be analysed: this then, in theory, is the simplest case. Unfortunately, in practice, the very identity of the particles introduces a difficulty: it is now impossible to distinguish, by considerations of velocity or by any other criterion, the α particles scattered in a given direction from the helium nuclei which are projected in that direction. This being so, equations (3) and (5) predict that the total number of particles in motion between θ and $\theta + \delta\theta$ to the axis of the beam is proportional to

$$(\cot \theta \operatorname{cosec}^2 \theta + \tan \theta \sec^2 \theta)\, \delta\theta \qquad \text{......(6)}.$$

As already indicated, the experimental results[a] did not at all conform to this prediction, but, because of the above-mentioned complication, it was not at the time easy to interpret them unambiguously in terms of the finite size and structure of the α particle. Since, in theory, this knowledge is an essential prerequisite for the complete explanation of anomalous scattering in any case, this attempt to establish a secure basis for such explanations must be said to have failed. The matter has been discussed at length only because it will be taken up again, from another point of view, at a later stage (p. 53).

Whilst, therefore, it cannot be claimed that the explanations of anomalous scattering current before 1929 were altogether consistent in themselves, yet, in so far as they did achieve a certain measure of success, it is worth while indicating their nature a little more fully. Two varieties of explanation are to be distinguished. Modifications in the law of force were expressed either analytically or by ascribing to the nucleus an abrupt boundary and thus a definite size and shape: as concerns the analytical expression it might be

[a] Rutherford and Chadwick, *Phil. Mag.* 4, 605, 1927.

grounded in a physical idea or it might be adopted empirically to fit the facts. The original explanations of anomalous scattering in hydrogen treated the matter from the second point of view, on the basis of a disk-like[a] or spheroidal[b] α particle. Explanations of anomalous scattering by heavier nuclei, on the other hand, were chiefly of the first type. They involved the assumption of a second term (inverse fourth power) in the law of force (Bieler),[c] or assumed electrical polarisation of the nucleus as the physical basis for an inverse fifth power term in the analytical expression (Debye and Hardmeier).[d] Even here, however, Rutherford and Chadwick[e] attempted an explanation of a more purely geometrical character. Clearly, when pushed to these extremes, the classical concept of the nucleus begins to appear a little less convincing.

§ 8. *Nuclear disintegration.* Very soon after the experiments of Moseley had resulted in the identification of atomic number and nuclear charge number, as above described, certain generalisations of Russell, Fajans and Soddy[f] were naturally interpreted on the nuclear hypothesis as showing that both the α particles and the primary[g] β particles emitted from radioactive substances were expelled from the nuclei of the atoms undergoing transformation. A similar conclusion concerning the origin of a large fraction[h] of the observed γ radiation became generally accepted about 1922.[i]

[a] Rutherford, *loc. cit.*; see also Darwin, *Phil. Mag.* 41, 486, 1921.

[b] Chadwick and Bieler, *Phil. Mag.* 42, 923, 1921.

[c] Bieler, *Proc. Camb. Phil. Soc.* 21, 686, 1923; *Proc. Roy. Soc.* 105, 434, 1924.

[d] Debye and Hardmeier, *Phys. Z.* 27, 196, 1926; Hardmeier, *ibid.* 28, 181, 1927.

[e] Rutherford and Chadwick, *Phil. Mag.* 50, 889, 1925.

[f] Rutherford, Chadwick and Ellis, *Radiations from Radioactive Substances*, 1930, p. 34.

[g] The data at that time merely indicated that β transformation of a radioactive atom involved the emission of one electron from the nucleus. Whether the general β radiation observed from matter in bulk included other electrons besides these "primary" β particles was left an open question. Subsequent experiments—and, most clearly, expansion chamber investigations with weak sources (Kinoshita, Kikuchi and Hagimoto, *Jap. J. Phys.* 4, 49, 1926; Feather, *Proc. Camb. Phil. Soc.* 25, 522, 1929) have shown that this is frequently the case (compare footnote, p. 4).

[h] Ideally the term "γ radiation" should be applied to this fraction only of the total high frequency radiation from disintegrating atoms.

[i] Ellis, *Proc. Roy. Soc.* 101, 1, 1922.

The acceptance of these conclusions led naturally to attempts to explain the ascertained facts of radioactivity in terms of nuclear structure conceived after the classical macroscopic model. Unfortunately, however, the facts themselves were not of very great variety. Mass and charge numbers were known for the nuclei in question, the energies of α, β and γ radiations had been determined in a majority of cases and the values of decay constants were well established. Concerning decay constants it was realised that these quantities represent the probabilities, per unit time, that the various nuclei will disintegrate. It was clear that these probabilities are quite independent of the times during which the nuclei have previously existed—apparently in complete stability. In the case of α disintegration, in which for each radioelement α particles of a single energy only were observed, an empirical relation appeared to connect decay constants with the energies of expulsion of the particles.[a] An attempt was made by Lindemann[b] to derive this relation on the basis of a mechanical interpretation of the probability of disintegration—as the probability of the occurrence of a particular (unstable) configuration in the course of the independent motions of N nuclear particles. Lindemann showed that disintegration constants of the right order of magnitude might be obtained on the assumption of quite reasonable values for N. Somewhat later this point of view was criticised by Campbell[c] who expressed cautiously the opinion that the chance nature of radioactive phenomena would eventually come to be regarded as fundamental, not standing in need of any mechanical explanation. It will be seen that this suggestion anticipates the modern outlook with some precision (cf. p. 49).

During the period 1913 to 1928 other nuclear models[d] were put forward chiefly with a view to calculating energies for α

[a] Geiger and Nuttall, *Phil. Mag.* 22, 613, 1911; *ibid.* 23, 439, 1912.

[b] Lindemann, *Phil. Mag.* 30, 560, 1915.

[c] Campbell, *The Structure of the Atom*, 1923, p. 52; see also Soddy, *Phil. Mag.* 18, 739, 1909; Rosseland, *Z. Physik*, 14, 173, 1923.

[d] It should be emphasised that quantum conditions, of the type applied to the outer atom by Bohr, are an essential feature of these models. That, however, does not disqualify them from consideration as "classical" structures (compare footnote, p. 34).

and γ emission which should be in sufficient agreement with experiment as to warrant some optimism regarding their significance. Thus Thibaud[a] proposed a model which had the merit of "explaining" most of the lines in the γ-ray spectrum of mesothorium 2: this postulated a number of protons[b] in energy levels under the action of a central positive core. Rutherford[c] achieved a large measure of numerical success in obtaining, by means of a model, values of the energies of the α particles emitted in all disintegrations: for this model neutral particles of mass number 4 were supposed to move in the outer nucleus under polarisation forces due to the field of a central core. Certain assumptions, however, were necessary regarding the actual mechanism of α emission which were of an arbitrary nature. Finally, Enskog[d] based a nuclear model essentially on the hypothesis of the existence of magnetic forces between nuclear particles in addition to forces of an electrostatic origin. In the complete calculations of α particle energies on this model the full validity of electromagnetic equations was assumed. Apart from other considerations, it is a little disconcerting to find that two entirely different modes of approach should each lead to fair numerical agreement with experimental data. The suspicion is that in each case the number of arbitrary constants at the disposal of the author was the chief cause of the success which he attained.

If, then, the nuclear models current in 1928 can be accorded but slight significance, there remains, from the classical viewpoint, a fundamental difficulty concerning experimental data first pointed out by Rutherford and Chadwick in 1925. This difficulty arises in a comparison of the scattering of α particles by uranium and their spontaneous emission in the α disintegration of that substance. From the classical point of view the observed energy of motion of an α particle spontaneously

[a] Thibaud, *Comptes rendus*, 181, 857, 1925.

[b] The term "proton" was introduced by Rutherford, in 1920, to describe a particle having mass number and charge number each equal to unity. In the free state (p. 12) it may be identified with the nucleus of the hydrogen atom of atomic mass very nearly equal to unity on the chemical scale.

[c] Rutherford, *Phil. Mag.* 4, 580, 1927.

[d] Enskog, *Z. Physik*, 45, 852, 1927; *ibid.* 52, 203, 1928.

emitted from a nucleus must in part be ascribed to the electrostatic repulsion experienced by it in its recession from the nucleus. To assume that the whole of its energy arises in this way is to conclude that the point of origin of the particle is as near as may be allowed to the centre of the nuclear structure. Now the α particles emitted spontaneously from uranium have an energy of about 4 million electron volts, whereas α particles of 7·5 million electron volts energy from radium C′ are scattered by a thin layer of uranium according to the detailed predictions of equation (2). Such of these particles as are scattered through approximately 180° are effectively brought to rest in the nuclear field and regain almost their full energy as they leave the field. At their closest approach to the centre of the nucleus they must obviously be within the distance from which the naturally emitted α particles are believed to be liberated. The results of the scattering experiments, therefore, suggest that even within this distance an approaching α particle is repelled by inverse square law forces due to a still more concentrated nucleus. Clearly, then, the classical interpretation suffers complete breakdown: α particles cannot be thought of as held in equilibrium in a region where the resultant force is one of repulsion—they cannot be thought of as permanently occupying that region at all if the whole of the charge of the nucleus is contained within a much smaller sphere.

So far nothing has been said regarding the interpretation of data relating to β disintegration. From the classical point of view very little can be said. The experiments of Ellis and Wooster[a] and Meitner and Orthmann[b] are unambiguous on the crucial question: in β disintegration the energies of individual β particles emitted from a radioactive preparation containing atoms of one kind only in general vary over a wide range[c]—from zero to a well-defined upper limit[d] characteristic of the radioelement in question—in some cases without there being emitted energy in any other recognisable

[a] Ellis and Wooster, *Proc. Roy. Soc.* 117, 109, 1927.
[b] Meitner and Orthmann, *Z. Physik*, 60, 143, 1930.
[c] Chadwick, *Verh. d. D. Phys. Ges.* 16, 383, 1914.
[d] Gurney, *Proc. Roy. Soc.* 109, 540, 1925.

form. If this conclusion is accepted as strictly correct it can only be inferred that initial (or final) nuclei are not all of exactly the same mass, although their remaining characteristics are single-valued. From the classical point of view it will be seen that the validity of the mass-energy conservation law is in question: the interpretation now current (p. 133) consists essentially in making *ad hoc* hypotheses to salvage that law; however successful it may be it is hardly one which might not have been proposed at a much earlier date.

From the first, experiments on artificial disintegration by α particle bombardment were discussed from the standpoint of the nuclear hypothesis. It was assumed that the disintegration particle—in all cases closely examined this appeared to be a proton—was expelled from the nucleus during the collision and the experiments of Blackett[a] showed that in the case of nitrogen, at least, the α particle was retained by the residual nucleus for a considerable period after the disintegration. Gradually, this type of capture disintegration, with the α particle permanently retained, was assumed as a basis of interpretation in all cases. It is clear from energy considerations alone that capture disintegration must be assumed whenever (as, for example, with aluminium) it is found that protons are liberated with kinetic energy greater than that of the α particles effective in producing the disintegration.[b]

In 1929, ten years after the first discovery of the effect, precise data concerning artificial disintegration were still somewhat scarce, but energy considerations of a general nature had already revealed difficulties of the type previously discussed (p. 40). The classical viewpoint was as follows: if an α particle is to be captured by a nucleus it must obviously penetrate to a region where the resultant force on it is attractive; if a proton escapes it must acquire energy in the repulsive field of the nucleus in moving from the place where the resultant force is zero to "infinity", and the energy with which it is observed cannot be less than this amount. Now

[a] Blackett, *Proc. Roy. Soc.* 107, 349, 1925.

[b] If such an "exothemic" change were possible without capture of the α particle it is to be supposed that it would take place spontaneously, which is assumed contrary to observation.

the analytical treatment[a] of the results concerning anomalous scattering allows of an estimate being made of the energy which the α particle must possess in the former case. For penetration within the attractive field of the aluminium nucleus it appeared that an α particle of roughly 8 million electron volts energy would be required. Direct experiment showed, however, that α particles of little more than half this energy might produce disintegration.[b] The only consideration which to some extent minimises this discrepancy is the recognition that the above-mentioned analytical treatment represents one only of the many attempts which have been made, from the "classical" standpoint, to explain the nature of anomalous scattering (p. 37). However, about the same time (1929), other difficulties arose which were not so easily disposed of. The most important, that provided by the results of Pose,[c] will be discussed in detail later (p. 158); less fundamental problems are involved in a difficulty which was first put in clear evidence by Rutherford and Chadwick.[d] By direct experiment with thin foils, these authors showed that the balance of kinetic energy in the disintegration of aluminium by α particles is far from constant. It was suggested that a discrepancy of the same type as that found in cases of radioactive β radiation might in fact be in question. Actually, a much more natural interpretation has since been found. More refined experiments have shown that discrete, rather than continuous, variations in reaction energy are involved and an explanation in terms of γ radiation has thus become possible. This, however, belongs to a later period; a full discussion will be found in chap. X. For the present a sufficiently large number of unsolved problems has been noted for it to be perfectly clear that the nuclear hypothesis requires something more than the background of classical physics to be of much use in the ordering of experimental data in this field.

[a] Debye and Hardmeier, *loc. cit.* (p. 38).
[b] Chadwick, *Phil. Mag.* 2, 1056, 1926.
[c] Pose, *Phys. Z.* 30, 780, 1929; *Z. Physik*, 64, 1, 1930.
[d] Rutherford and Chadwick, *Proc. Camb. Phil. Soc.* 25, 186, 1929.

CHAPTER III

THE NEED FOR A NEW POINT OF VIEW: INTERPRETATION IN TERMS OF WAVE MECHANICS

§ 9. *Difficulties of a general nature.* Already a number of examples has been given of nuclear phenomena which have not been satisfactorily interpreted within the ideological framework of classical physics. Before proceeding to the fundamental ideas which form the basis of a much more successful interpretative scheme—the so-called wave mechanics—it will be well to mention very briefly a few of the difficulties which were beginning to be apparent, elsewhere than in the nuclear domain, in the period 1920–25. These difficulties, more than those which have already been noticed, point the way to the general solution.

First of all there is a logical difficulty. The old quantum theory of the atom, although very successful in many respects, was formally unsound—or, at the least, incomplete—in that Planck's constant, h, was introduced twice, quite arbitrarily on each occasion, in the course of the calculations. There was obvious need of an explanation why this procedure should result in success; why, in the quantisation of momentum and in the relation between radiation frequency and energy, apparently the same natural constant should be involved. The old quantum theory provided no such explanation.

Then there were the experimental difficulties. As the data of optical spectrum analysis became increasingly more complex, ever greater difficulty was continually experienced in retaining a systematisation which should be physically consistent with the mechanical properties of the atom model generally adopted. These difficulties are mostly matters of detail which cannot be referred to here,[a] but there were also

[a] Cf. Andrade, *The Structure of the Atom*, 1927, chaps. XV and XIX; Born, *Atomic Physics*, 1935, chap. V.

others not similarly obscured in the data of a complicated subject. On the one hand, experiments on the reflection or scattering of electrons from metallic crystals had provided evidence of a spatial distribution of scattered particles very suggestive of a diffraction effect[a] and, on the other, the lack of temperature dependence of the "cold electron emission" from clean metal surfaces[b] appeared quite incomprehensible on the basis of accepted ideas. Somewhat more complex, it seemed, but equally incomprehensible, was the Ramsauer effect—the unexpected rise and fall of atomic collision cross section with increasing velocity, for slow electrons in the inert gases.[c] These were the first observed manifestations of a duality in the properties of matter very similar to that which had long been recognised in the properties of light or radiation. Moreover, this analogy is itself the starting-point of the more successful method of interpretation.[d]

Here it may not be without point to interpolate that, whilst it is the avowed object of any physical theory to correlate observed phenomena (events), this cannot readily be done without the assumption of "enduring objects" of some kind. In our introduction to the classical viewpoint (§ 6) we have already, in effect, mentioned two types of enduring object which derive naturally from common-sense experience: material particles and groups of waves. In the nineteenth century the small scale phenomena of optics and molecular physics were described—by an obvious, but, it must be admitted, arbitrary choice—the former entirely on a wave basis, the latter exclusively in terms of material particles. The insufficiency of this one-sided description became evident in the optical domain nearly thirty years earlier than in the sphere of molecular physics. Now, however, duality is seen to be an essential feature of both; of this duality wave mechanics

[a] Davisson and Kunsman, *Phys. Rev.* 22, 242, 1923; Davisson and Germer, *ibid.* 30, 705, 1927.

[b] Lilienfeld, *Phys. Z.* 23, 506, 1922; Gossling, *Phil. Mag.* 1, 609, 1926; Millikan and Eyring, *Phys. Rev.* 27, 51, 1926; de Bruyne, *ibid.* 35, 172, 1930.

[c] Ramsauer, *Ann. Physik*, 64, 513; *ibid.* 66, 546, 1921; *ibid.* 72, 345, 1923; Ramsauer and Kollath, *ibid.* 3, 536, 1929; *ibid.* 4, 91, 1930.

[d] de Broglie (L.), Thèse, Paris, 1924; *Ann. Physique*, 3, 22, 1925; Schrödinger, *Ann. Physik*, 79, 361, 1926.

provides the formal expression. As a result, whilst it appears best in each case to describe events in terms of particles, in each case, also, a wave description is clearly necessary to furnish the theoretical counterpart of the conceptual objects which are supposed to "endure". On the older theory the workings of a physical system between observed events were discussed in terms of particles of mass $m_1, m_2, \ldots$ moving with velocities $v_1, v_2, \ldots$; now wave-lengths $\lambda_1, \lambda_2, \ldots$ must be introduced.

Since we are not concerned with a formal exposition of the subject we may say that Planck's constant, h, is involved merely in the relation between these two methods of treatment. Instead of a particle moving in a given direction with velocity v (though with position completely undefined) we have a plane wave motion of wave-length λ given by

$$mv\lambda = h \qquad \ldots\ldots(7).$$

Definition of position, if this is necessary, is represented by superposing wave motions of neighbouring wave-lengths in such a way as to produce a group of waves. Necessarily the group velocity in this case must also be v. The phase velocity V is given by

$$vV = c^2 \qquad \ldots\ldots(8),$$

where c is the velocity of light.[a] Clearly, this specification of position is not precise, but again there is a straightforward connection between the two points of view: the probability of the instantaneous existence of the particle in a particular element of volume is proportional to the square of the wave amplitude in the volume element concerned. In this paragraph, it will be observed, we have been dealing merely with a single particle and the corresponding wave treatment of its motion. To a large extent such considerations suffice for a general discussion of the nuclear phenomena with which we shall be concerned, but it should also be remarked, to avoid misconstruction, that when two or more particles are under

[a] On the older theory, of course, the total energy of the particle, at any instant of observation, is given by mc^2 at that instant. We may say, therefore, that Planck's constant is involved—as before—in relating energy to wave frequency: cf. equations (7) and (8), whence $mc^2 = h(V/\lambda)$.

discussion a complication arises. The wave "processes", then, require more than three spatial dimensions for their treatment. Between the "phase space", which has to be employed, and the space of experience only formal relations subsist. Strictly speaking, therefore, the three dimensions which are sufficient[a] for single particle problems should also be regarded as of phase space, only the formal relations with actual space in this case are particularly direct.

It has been stated that a single particle treatment is frequently sufficient for the problems with which we shall be concerned. Obviously, however, no strictly single particle problem is of any immediate interest, rather must our statement be interpreted as follows: it is very frequently sufficiently accurate to regard the interaction between two particles, one light and one heavy, or between a single particle and matter in bulk, as the interaction between an isolated particle and a field of force. If the latter is described in terms of the potential energy of the particle from point to point in the field all the data are available for the calculation of velocities (or wave-lengths) at any stage of the motion, once the total energy of the particle is given. This, again, is sufficient for simple considerations.

One important difference between the new treatment and that of classical mechanics may be noticed at this stage: it is that sufficiently narrow regions of imaginary velocity (negative kinetic energy) are not, according to wave theory, necessarily impassable; finite "barriers of potential energy" are not absolutely prohibitive.[b] The treatment of radioactive disintegration—and of cold electron emission, already mentioned—is based upon this result.

§ 10. *Scattering and disintegration.* Let us return to the consideration of the scattering of α particles by (radioactive) uranium (p. 40). Fig. 1 may be taken to represent the potential energy of an α particle at various (radial) distances from the centre of a uranium nucleus. Over most of the

[a] If it be assumed that the particle is entirely devoid of spin.

[b] Nordheim, *Z. Physik*, 46, 833, 1928; Oppenheimer (J. R.), *Phys. Rev.* 31, 66, 1928.

region, with sufficient accuracy, this energy is given by $2 \times 92 \times e^2/r$ (distances small compared with the mean distances of the innermost atomic electrons), but near the centre, clearly, a negative term must be included to represent the fact of approximate stability.[a] On the same diagram horizontal lines E_0 and E_1 indicate, on a common energy scale, the kinetic energies of the α particles emitted by uranium and thorium C′, respectively. Particles of the latter energy are supposed used in the scattering experiments.

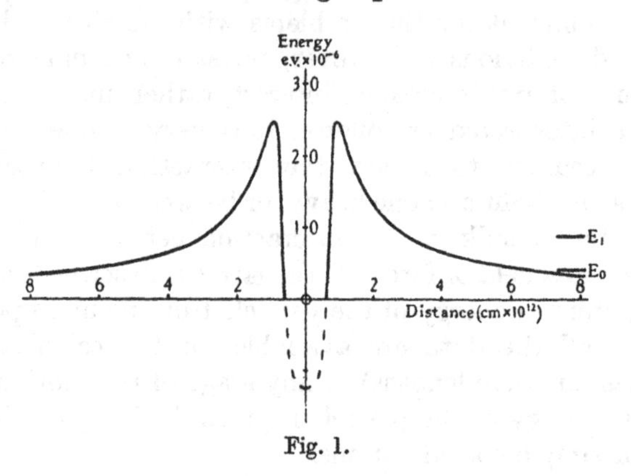

Fig. 1.

In adopting the new point of view it is necessary first of all to enquire what scattering would be expected in a strictly inverse square law field. As already stated, classical considerations in this case lead to the angular distribution function of equation (2). The calculations of Wentzel[b] and Mott[c] showed that the same distribution law was valid[d] on the basis of wave mechanics. This is an important conclusion. It immediately confirms the previous interpretation of the experi-

[a] Although little in detail is known about this additional term, the form of the resultant curve given in the figure is that generally adopted; cf. Gamow, *Structure of Atomic Nuclei and Nuclear Transformations*, 1936, chap. V.

[b] Wentzel, *Z. Physik*, 40, 590, 1926.

[c] Mott, *Proc. Roy. Soc.* 118, 542, 1928.

[d] In spite of this fact there is an important difference between the two viewpoints. According to classical ideas there is a one-to-one correspondence between angle of scattering and impact parameter—this is the statistical basis for the distribution law—but in wave mechanics no such exact correlation is contemplated.

ments in question—up to the height[a] E_1 the barrier of potential energy between an α particle and the uranium nucleus is of the standard form. But it requires, also, the following qualification (which does not arise in the classical treatment): in so far as actually the general features of Fig. 1 must be accepted as correct (a barrier with finite height), the chance of penetrating the barrier at the height E_1, though negligible for a single encounter, is nevertheless presumably finite. On any simple supposition this finite chance is smaller still at the height E_0. A consideration of the long lifetime of uranium, which emits α particles of energy E_0 spontaneously, shows that actually for this energy the chance[b] must be of the order of 10^{-39}.[c] There is no obvious inconsistency in these conclusions once the finite transparency of potential barriers is assumed. The qualitative and quantitative application of this assumption to the explanation of radioactive disintegration was made independently by Gamow[d] and by Gurney and Condon,[e] first in 1928. We shall return to its further consideration in § 24.

In discussing the scattering of α particles by lighter nuclei the only general difference to be noted is in the height of the peak of the potential barrier. The fact of "anomalous" scattering, which becomes more pronounced, for α particles of a given energy, the smaller the atomic number of the element investigated (p. 36), is clearly to be explained, on the new theory, by the circumstance that the particle energy and the energy corresponding to the summit of the barrier

[a] In employing this phraseology it should be remembered that conditions are much simplified by the assumption of spherical symmetry. In the general case space of four dimensions would be required for the representation of the potential barrier surrounding a nucleus.

[b] In discussing the disintegration of uranium, strictly the potential barrier between an α particle and the subsequent nucleus (UX_1) should be employed, also, in scattering and in disintegration, the energy which is taken should be the energy of relative motion of the two particles, not that of the α particle alone. The main results stand, however, after these corrections have been made.

[c] The uranium α particle may be regarded as making about 10^{39} collisions with the inner wall of the barrier during the average life of the nucleus, whilst, even with thorium C′, the escaping particle will previously have made, on the average, roughly 10^{16} such "unsuccessful" collisions.

[d] Gamow, *Z. Physik*, 51, 204, 1928.

[e] Gurney and Condon, *Nature*, 122, 439, 1928; *Phys. Rev.* 33, 127, 1929.

are for these elements comparable in magnitude. Penetration of the barrier occurs to an appreciable extent, with the result that quite a large amount of scattering takes place in the inner attractive field and must be added to the normal scattering occurring in the region outside the nucleus. According to the methods of wave mechanics this addition is made by considering the superposition of scattered "wavelets" and so proceeding to the square of the resultant wave amplitude at any point, as already described. The problem has been discussed theoretically by Beck[a] and subsequently by a large number of other investigators. Since detailed predictions in all cases depend upon a knowledge of the form of the potential barrier within the summit, it is clear that such predictions cannot, in general, be expected;[b] however, the broad method of calculation may be indicated here because of its physical significance. Three steps are involved: the plane wave representing a unidirectional beam of particles is decomposed into spherical harmonic components of successive orders, the scattering of each component is considered separately and, finally, the corresponding "scattered waves" are recombined. The initial decomposition will be seen to correspond to a classification of the incident particles according to angular momentum about the centre of force (or according to the impact parameter, p, of classical mechanics) and the general result, that anomalous scattering, when appreciable, is always to be attributed to the behaviour of the spherical harmonic components of lowest order, is evidently the counterpart of the classical statement that the approach to the nucleus is always closest when p is least, for α particles of a given energy. As we have seen, the fact of anomalous scattering, according to present ideas, is held to imply that an appreciable fraction of the deflected particles traverse the attractive field of the nucleus in the process of scattering. These particles may be said to penetrate to some extent

[a] Beck, *Z. Physik*, 62, 331, 1930.

[b] Although, ideally, the reverse process—the deduction of the inner field from the observed scattering—may be possible in simple cases; cf. Taylor (H. M.), *Proc. Roy. Soc.* 134, 103, 1931; *ibid.* 136, 605, 1932; Wheeler, *Phys. Rev.* 45, 746, 1934; Mohr and Pringle, *Nature*, 137, 865, 1936.

within the nucleus. Now just such assumptions have already been made, on a classical basis, concerning the α particles which are effective in producing the artificial disintegration of light nuclei. According to wave mechanics, also, whenever there is anomalous scattering there is, in addition, the possibility (if energy conditions be fulfilled) that such disintegrations will occur; again the same set of particles is responsible for the two phenomena. A wave-mechanical theory of artificial disintegration was first given by Gamow[a] in 1928. At that time, however, only a small amount of the present data was available for comparison and it was not difficult to establish the general result, given by experiment, that the probability of disintegration, for a given energy of α particle, becomes negligibly small when the atomic number of the bombarded substance is even moderately high (say $Z > 20$). Gurney[b] made the first suggestion which introduced any complication into the discussion. He pointed out that resonance phenomena might be expected to occur when one or more of the virtual proper energies for an (additional) α particle ir the bombarded nucleus happened to lie in the energy range covered by the bombarding particles. In such a case the probability of disintegration would no longer increase monotonically with the energy of the α particle.[c] In actual fact, an effect of this kind was, about that time, in process of elucidation by Pose[d] (cf. pp. 43, 158). With the proof of the reality of these resonance phenomena many investigators[e] joined in the theoretical discussion[f] of the subject. Two general conclusions may be given. First, if proper energies are to be calculated from the theoretical side, a

[a] Gamow, *Z. Physik*, 52, 510, 1928.

[b] Gurney, *Nature*, 123, 565, 1929.

[c] This probability is greatest when the energies of incident particle and virtual level are equal, but the magnitude of the effect which is observed clearly depends on the "breadth" of the resonance level, as well as upon the effective transparency of the barrier at its centre.

[d] Pose, *Phys. Z.* 30, 780, 1929.

[e] E.g. Beck, *Z. Physik*, 64, 22, 1930; Mott, *Proc. Roy. Soc.* 133, 228, 1931.

[f] We say that at certain (resonance) energies the α particle has an abnormally high chance of penetrating the nuclear structure because, in the wave treatment, it is shown that in the steady state a standing wave of great intensity is obtained "inside the barrier" for just the wave-lengths which correspond to these energies.

detailed knowledge of the inner field of the nucleus must be assumed and, secondly, since the sharpness of resonance is found to depend on the azimuthal quantum number of the virtual level, the possibility arises of deducing not only proper energies, but also quantum numbers, from the experimental data. Up to the present, however, progress in this direction has not been very considerable. Again, whilst anomalous scattering as the result of resonance penetration has been discussed theoretically,[a] evidence is still lacking which would establish this phenomenon in fact.

So far in this discussion of artificial disintegration one very important point has scarcely been mentioned. We have referred to penetration within the nucleus and to disintegration indiscriminately, almost as if the two terms were synonymous. Actually, it is clear that this is far from the case. Evidently the requisite energy must in fact be available (this condition has already been given), but even then the probability of disintegration, assuming penetration, may vary widely from one case to another. Two determining factors are to be distinguished: there is the change of spin which accompanies the process and the probability factor which describes simply the chance of emission (within the time which is available) of the disintegration particle from the temporarily unstable nuclear system. With the first consideration we shall not deal further here—it has been discussed in a number of places, as, for example, by Goldhaber[b] —but the latter point may be very briefly treated. It amounts merely to this: the disintegration particle must be thought of as leaving the nucleus against a barrier of potential energy appropriate to its nature.[c] The success of the earlier calculations concerning the emission of protons under α particle bombardment is to be traced to the fact that the potential barrier for the proton, in any case, is so much lower than that for the α particle that, unless strongly endothermic changes are in question, the latter determines disintegration probabilities rather than the former. But it should be

[a] Mott, *loc. cit.* [b] Goldhaber, *Proc. Camb. Phil. Soc.* 30, 561, 1934.

[c] Beck, *loc. cit.* Concerning the emission of neutrons, see p. 185.

pointed out, first, that resonance effects in respect of the emitted particles may sometimes occur and, secondly, that the disintegrations which are produced by neutrons must certainly be discussed (cf. p. 178) in terms of the potential barriers appropriate to the charged particles which are ejected.

§ 11. *Particles of one kind* ("*identical*" *particles*). It has already been stated that the predictions alike of classical mechanics and of the wave mechanics are represented precisely by the Rutherford scattering formulae (pp. 33, 35) in the physically important case when the interaction between the moving particle and the scattering centre is of the inverse square law type. In actual fact one marked exception must be made. As was first pointed out by Mott,[a] the statement is no longer true for collisions between particles of a single kind, thus a different distribution law is to be expected for the scattering of protons in hydrogen, of α particles in helium, or for the electronic scattering of fast electrons. It may be stated at once that experimental investigations have confirmed the predictions of the wave mechanical treatment in all three cases.[b] Here it remains to indicate the reason for the differences in question.

In discussing the scattering of α particles in helium on a classical basis we have previously (p. 37) referred to the experimental difficulty that the "scattered" and "projected" particles travelling in a given direction (with respect to the axis of a collimated incident beam) have the same velocity. There being no way of examining the two sets of particles separately, their joint number was originally taken to represent the sum of the scattering through the experimentally determined angle and through its complement. Essentially, at the basis of the wave mechanical treatment of the problem, is the denial of the validity of the distinction (even in thought) between "scattered" and "projected" particles, when the

[a] Mott, *Proc. Roy. Soc.* 126, 259, 1930; see also Oppenheimer, *Phys. Rev.* 32, 361, 1928.

[b] Gerthsen, *Ann. Physik*, 9, 769, 1931; Chadwick, *Proc. Roy. Soc.* 128, 114, 1930; Blackett and Champion, *ibid.* 130, 380, 1931; Williams, *ibid.* 128, 459, 1930.

two are of one kind. The number of particles which is observed in a given direction is then given not by adding intensities, but by combining wave amplitudes (taking count of phase differences, of course) and calculating from the square of the resultant amplitude, as already discussed. Clearly, accepting a theory of this type, we might expect diffraction effects, numbers of particles which do not vary monotonically with the angle of observation. This is precisely what is predicted—and what has also been found—with the details of the phenomenon depending upon the spin of the particles and the type of statistics applicable to them. For α particles in helium (spin quantum number, I, $=0$; Einstein-Bose statistics) the classical result is modified by replacing the factor (6), p. 37, by

$$\left\{\cot\theta\,\mathrm{cosec}^2\theta+\tan\theta\sec^2\theta\right.$$
$$\left.+2\,\mathrm{cosec}\,\theta\sec\theta\cos\left(\frac{8\pi e^2}{hv}.\log\tan^2\theta\right)\right\}\delta\theta \quad\ldots\ldots(9),$$

whilst for protons in hydrogen ($I=\frac{1}{2}$, Fermi-Dirac statistics) the corresponding factor is

$$\left\{\cot\theta\,\mathrm{cosec}^2\theta+\tan\theta\sec^2\theta\right.$$
$$\left.-\mathrm{cosec}\,\theta\sec\theta\cos\left(\frac{2\pi e^2}{hv}.\log\tan^2\theta\right)\right\}\delta\theta \quad\ldots\ldots(10).$$

In each case v is the initial velocity of the particle which is scattered. It will be seen, first of all, that the scattering in these cases is, respectively, twice the classical amount and half that amount for $\theta=\frac{\pi}{4}$, independently of the velocity of the particles, and, secondly, that diffraction effects, represented by the third terms in (9) and (10), are more pronounced the smaller the velocity. The third term in (9), for example, is zero for the angles $\theta_1, \theta_2, \ldots, \theta_n, \ldots$, where

$$\tan\theta_n=e^{-(2n-1)\beta v} \quad\ldots\ldots(11),$$

with

$$\beta=h/32e^2 \quad\ldots\ldots(12)^{a}:$$

[a] Two different usages for e are represented in (11) and (12), but it is not thought that any confusion will thereby arise.

clearly, as v is made smaller, these angles become closer together. The chief experimental difficulty in establishing the effect with α particles and protons was the difficulty of employing particles of sufficiently low velocity. In any case, as already described, the disturbing effect of anomalous scattering enters if the velocity of the particles is too high.

In order to explain the difference attributed to the spin of the particles the following observations are relevant (as it is the important one, we shall take the case for which $I=\frac{1}{2}$). Consider a number of similar collisions after each of which equal particles move at angles θ and $\frac{\pi}{2}-\theta$, respectively, to the initial direction. Consider the particles moving in the former (θ) direction. The axes of spin of any two such particles may be parallel or anti-parallel, actually, of course, with the former contingency intrinsically the more probable. The important point for the present discussion, however, is not the difference in *a priori* probabilities, but rather the fact that only when the spin axes are parallel are the particles completely identical for the purposes of the theory. The case of equal particles with non-zero angular momentum is thus seen to be intermediate between that of equal particles devoid of spin and the case of unequal particles. The exact form of the expression (10) comes from combining the appropriate distribution functions with due regard to prior probabilities.

CHAPTER IV

ELEMENTARY PARTICLES: NUCLEAR STRUCTURE

§ 12. *Structural units.* For three quarters of a century the concept atom-of-matter was employed in discussions of physical phenomena without its becoming necessary to particularise further and endow the atom with a structure of its own. Nineteenth-century theoretical physics was eminently successful in its use of this simple concept, experiment providing abundant confirmation of the significance of results attained by that use. The concept atom-nucleus has had a very different history. Its general acceptance was followed almost immediately by the realisation that a considerable degree of complexity must be assigned to the nucleus. Radioactivity, as a property of heavy nuclei, and susceptibility to artificial disintegration, as a property of light nuclei, can only be interpreted in terms of a nucleus possessing internal structure (see § 8).

At first[a] it was natural to suppose that three different structural units were involved: α particles and β particles, since these are emitted spontaneously from radioactive substances, and protons, which may be ejected by bombarding certain light elements with α particles.[b] It was assumed that particles of all three types might exist as nuclear constituents possessing almost complete individuality. Moreover, from this point of view, the α particle itself (the helium nucleus of mass number 4) could be assumed constituted of four protons and two electrons; it came to be regarded, therefore, as a secondary unit, albeit a most important one, for nuclear

[a] Cf. Gehrke, *Verh. d. D. Phys. Ges.* 21, 779, 1919; Wolff, *Ann. Physik*, 60, 685, 1919; van den Broek, *Phys. Z.* 21, 337, 1920; *ibid.* 22, 164, 1921; Rutherford, *Proc. Roy. Soc.* 97, 374, 1920; Beck, *Z. Physik*, 47, 407; *ibid.* 50, 548, 1928.

[b] An entirely different starting-point for considerations concerning nuclear structure has been adopted by Harkins and followed up in a number of papers: see, for instance, *Phil. Mag.* 42, 305, 1921.

structure. It is worthy of note that at this early stage it provided the single direct proof of a community of structure as between the lightest and the heaviest nuclei: an α particle might be incorporated in an already existing light nucleus in the process of capture disintegration (p. 42) or it might be emitted spontaneously when a heavy nucleus disintegrated. Only from one standpoint was the position less satisfactory: the total numbers of electrons and protons might be definite for a given nucleus—no alternative being possible if mass and charge numbers were to be satisfied[a]—but the extent to which these primary constituents occurred grouped together as α units remained entirely indeterminate.[b] Generally, considerations of simplicity, involving the assumption that the total number of nuclear particles should be a minimum, were invoked to complete the scheme. Then never more than three free protons had to be assumed for any known nucleus. For the heavier nuclei, however, free electrons, up to 28 in number, were necessary from this point of view.

On any assumption concerning nuclear structure one important consideration must be satisfied: if the final nucleus is to be stable its mass must be less than the sum of the masses of its constituents as determined in the free state. This "mass defect",[c] Δm, is a measure of the energy of binding, Δmc^2,[d] of the nuclear particles; it represents the minimum work which must be done if the nucleus is to be entirely disrupted into its component parts. It cannot be too strongly emphasised, however, that in our present state of ignorance there is no sense in referring to the mass defect of a given nucleus as a

[a] The assumption—universally accepted—that the number of protons in the nucleus is given by the mass number defined with reference to oxygen (footnote, p. 12) strictly does admit of alternative. Mass numbers defined with reference to hydrogen differ by unity from the above for $A \sim 130$; that they have not been more generally adopted may be traced to the corresponding uncertainty of this definition both when $A \sim 65$ and when $A \sim 195$.

[b] Brösslera, *Rev. Chim.* 1, 42, 74, 1921; Ono, *Proc. Math. Phys. Soc. Japan*, 8, 76, 1926; Rutgers, *Nature*, 129, 361, 1932.

[c] Swinne, *Phys. Z.* 14, 145, 1913; Harkins and Wilson, *J. Amer. Chem. Soc.* 37, 1367, 1915.

[d] When the value of the mass defect is given on the usual atomic scale ($O^{16} = 16$) and the binding energy is expressed in electron volts, the conversion factor c^2 is, numerically, $9{\cdot}3 \times 10^8$ electron volts per mass unit.

uniquely specified quantity which may be obtained once and for all by a suitable experiment. All that is possible is that the exact masses of nuclei shall be determined with increasing precision; the mass defect assigned to each, on the basis of these determinations, will depend upon the choice of building units for the nuclear model. For any model to be acceptable this choice must result in positive values for the mass defects of stable nuclei. Those calculated[a] for a simple electron-proton model are shown diagrammatically in Fig. 2. It will be

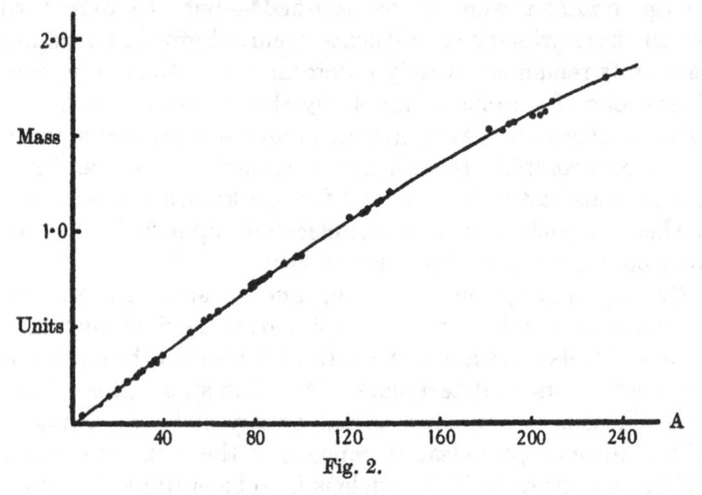

Fig. 2.

seen at once that the above condition is amply satisfied in this case. If α particles be included as secondary units, mass defects become considerably smaller throughout, the difference for each nucleus being just so many times $28{\cdot}6 \times 10^{-3}$ mass units, the mass defect for the α particle in terms of electrons and protons, as there are α particles in its structure. Fig. 3 represents the results when the maximum possible number of α particles is assumed in each case. On this hypothesis, also, positive mass defects are universal; on either assumption, it will be noted, the heavy radioelements are found to be stable as against complete nuclear disruption. On

[a] For the purposes of this calculation the masses given in Tables 4, 5 and 6 (pp. 78–80) are employed.

the other hand they are definitely unstable as regards certain types of disintegration. This fact suggests a closer analysis of the situation. Obviously any choice of structural units is unsatisfactory which would make the hypothetical change

$$X \rightarrow Y + \pi \qquad \text{......(13)},$$

from one stable nucleus X to another Y, with π as one of the common constituents of these nuclei, appear to be exo-

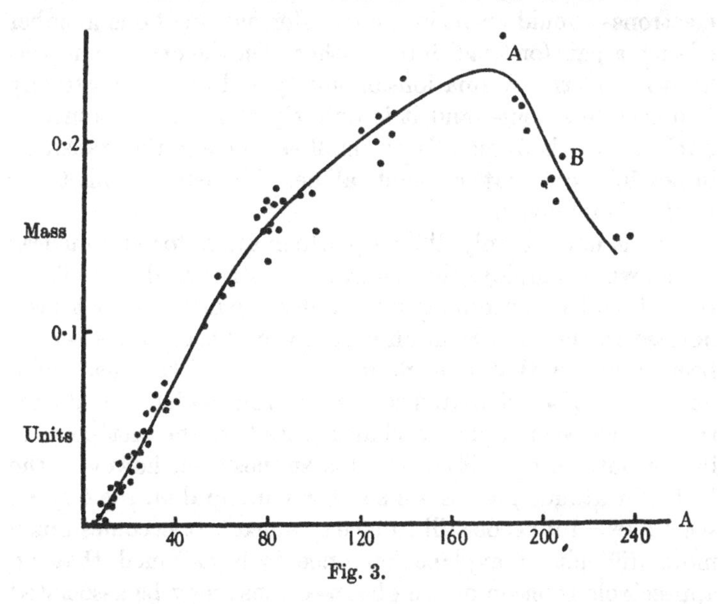

Fig. 3.

thermic. The general trend of the smooth curves of Figs. 2 and 3 suggests that the proton and the α particle are satisfactory units from this point of view. Where the trend of the curve suggests the contrary (Fig. 3, *AB*), a detailed examination of the masses of all nuclei reaffirms this conclusion; except in the region of the heavy radioelements, unit increase in the number of nuclear protons (or α particles) is invariably associated with the disappearance of nuclear mass. For the heavy radioelements, moreover, Fig. 2 suggests stability as against proton disintegration and Fig. 3 instability for α particle disintegration, in complete accord with the facts which any views concerning nuclear structure must seek to explain.

If, however, the particle π in (13) be the electron, difficulties at once arise. At present about 50 examples of isobaric pairs and triads are known amongst the stable nuclei, that is 50 cases in which two (or three) nuclei of different charge number have the same mass number. In some cases the emission of one electron—in most cases the emission of two electrons—would result in the transformation of one member of such a pair (or triad) into another. On the other hand it is known that transformations of this type do not occur to any appreciable extent—and it is unlikely that the difference in nuclear mass is frequently so small as to make the transition impossible on that account alone. Evidently something further is involved.

This is not the only difficulty which arises for any nuclear model which employs the electron as a structural unit. There is a difficulty concerning mass defects which has not been noticed in the discussion already given. There it has tacitly been assumed that a negligible fraction of the mass of a nucleus is to be attributed to the electrons which it contains: these have been supposed characterised by the rest mass as in the outer atom. Even on this supposition, however, the fact that atomic masses are so nearly integral on the oxygen scale ($O^{16} = 16$) needs still to be explained.[a] It becomes much more difficult of explanation once it is assumed that an appreciable fraction of the observed mass may be associated with electrons in rapid motion. Similarly, on that assumption, also, the identification of the nuclear mass number with the number of protons in the nucleus (p. 57) appears to rest on less plausible assumptions. Now this necessity of assigning high speeds to nuclear electrons, if such exist, has always characterised any quantum theory interpretation of the phenomena. On the original Bohr theory[b] the quantisation of

[a] It is sometimes suggested that an explanation based on the assumption of the maximum possible number of α particles in the nucleus is the most natural here. To complete the explanation it is necessary then merely to assume that the binding energy per α particle is roughly constant for all nuclei.

[b] Cf. Enskog, *Z. Physik*, 45, 852, 1927.

electron motion within the restricted space of the nucleus, and, from the standpoint of wave mechanics,[a] the application of the uncertainty relation to this problem, both lead to essentially the same result: nuclear electrons must be regarded as possessing velocities so high that they do, in fact, contribute appreciably[b] to the mass of the nucleus. This difficulty was fully recognised long before a natural way of escape was provided by new discoveries of experimental fact. So long as it was accepted, it was realised that only a completely relativistic treatment of the problem could have any chance of success. This, in itself, was for the time being sufficiently discouraging for most theorists (see, however, § 28, p. 137).

The new experimental facts, above referred to, are those which in 1932 led Chadwick[c] to conclude that in certain cases of artificial disintegration produced by α particle bombardment an uncharged particle ("neutron"; $A=1$, $Z=0$) is emitted from the nucleus transformed. Soon afterwards the experiments of Feather[d] provided evidence which was at once interpreted in terms of the disintegration of nuclei by capture of this particle. Twelve years previously Rutherford[e] had speculated concerning its existence and indicated its possible importance as a nuclear constituent; now Heisenberg,[f] in particular, proceeded to develop a nuclear model employing protons and neutrons as structural units. On this scheme it is still possible to retain the α particle as a secondary unit (formed of two protons and two neutrons), but it is to be noticed that the necessity of including the electron is entirely removed. To that extent a natural way of escape was provided from the particular difficulties elaborated above, but it remained to be seen whether other difficulties of equal stubbornness might not thereby be raised. The most obvious doubt concerns the description of the process of β disintegration

[a] Cf. Gamow, *Constitution of Atomic Nuclei and Radioactivity*, p. 2, 1931.
[b] Say 10^{-2} to 10^{-1} mass unit per electron.
[c] Chadwick, *Nature*, 129, 312, 1932; *Proc. Roy. Soc.* 136, 692, 1932.
[d] Feather, *Proc. Roy. Soc.* 136, 709, 1932; *Nature*, 130, 237, 1932.
[e] Rutherford, *Proc. Roy. Soc.* 97, 374, 1920.
[f] Heisenberg, *Z. Physik*, 77, 1, 1932.

from this point of view. Clearly it must be thought of as involving the transformation of a neutron into a proton within the nucleus. There were two reasons why this description was not at once rejected. In the first place, Meitner[a] had long previously suggested a mechanism for the emission involving the intranuclear transformation of neutral particles (α' particles, of mass number 4), on the basis of series regularities, and, secondly, about this time (1932) an entirely different, but quite clear-cut, example was discovered of a unit[b] particle being emitted from a nucleus, in the course of disintegration, which could not possibly have existed previously as such within the nucleus. The distintegration of lithium by proton bombardment was found[c] to result in the production of two α particles for each lithium nucleus disintegrated: obviously both these α particles could not have had separate existence in the original nucleus (p. 189). In fact the neutron-proton model was retained—and the description of β disintegration originally proposed in terms of it remains the orthodox description of that transformation process at the present time. If the attempt had been made to include as individual constituents of nuclei all those particles which are spontaneously, or by conditioned transformation, emitted[d]—and those which are captured by heavier nuclei in transformations of the latter type—then very quickly the positron (§ 26), the deuteron ($A = 2$, $Z = 1$) and two particles of mass number 3 (having $Z = 1$ and $Z = 2$, respectively) would have been added to the previous list, to the complete confusion of theory.[e] As it is, it is general to suppose that the particles in question assume individuality only at the moment of emission, whilst, for many purposes, this supposition is made in the case of α particles, also. If it is not made in this case (cf. chaps. VII

[a] Meitner, *Z. Physik*, 4, 146, 1921.

[b] This term is here used to denote a particle belonging to one of the types regarded as primary or secondary nuclear building units.

[c] Cockcroft and Walton, *Proc. Roy. Soc.* 137, 229, 1932.

[d] In this connection a particle is said to be emitted when, in a given transformation, it is the resultant of smaller (or smallest) mass.

[e] Gamow (*Nature*, 135, 858, 1935) has argued for the inclusion of a hypothetical negative proton, despite its non-occurrence in the transformations hitherto studied.

and X) and the individual existence of the maximum number of nuclear α particles is postulated as before (p. 57) it is worth noting that this number is now considerably less, for the heaviest nuclei, than the number previously given on the

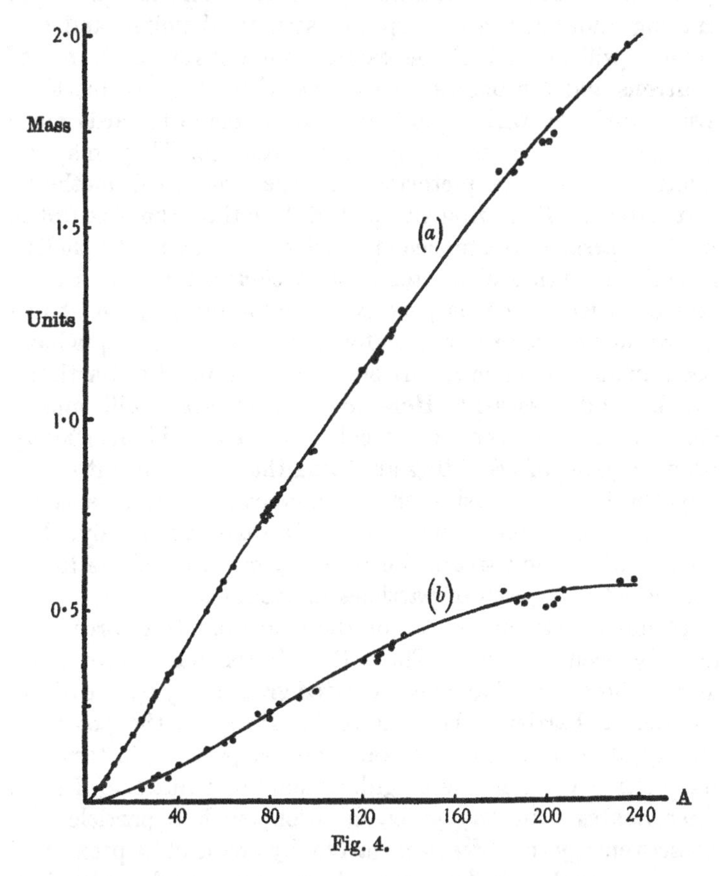

Fig. 4.

electron-proton model.[a] Thus, with neutrons and protons as constituents, the thorium nucleus ($A=232$, $Z=90$) contains 45 α particles, instead of 58 on the old model. Fig. 4, (*a*) and (*b*), gives the mass defects of nuclei regarded as constituted of neutrons and protons, (*a*) without and (*b*) with the assump-

[a] Iwanenko, *Comptes rendus*, 195, 439, 1932.

tion of (the maximum number of) α particles as secondary units. The smooth curves (*a*) and (*b*) may be compared with those of Figs. 2 and 3.

§ 13. *Structural patterns: laws of force between nuclear particles.* Having adopted the appropriate structural units—and discussion will henceforth be carried out entirely in terms of neutrons and protons as nuclear constituents—two methods are possible by which progress may be made towards more definite ideas concerning nuclear structure. They may be referred to as the pictorial and the analytical methods, respectively. Following the pictorial method the attempt is made to arrange the nuclear particles in a system of "shells" having saturation characteristics so chosen that the known stable nuclei appear as permitted configurations, whilst unknown nuclei are excluded as forbidden. Such attempts have been made by several writers, for example by Bartlett,[a] Landé[b] and Elsasser.[c] Here mention of them will suffice, since, in theory, the analytical method should eventually prove capable of providing anything that is significant in the hypothesis of successive shells; moreover, it is essentially quantitative whereas the former is qualitative only. The analytical method treats the problem in terms of the forces acting between pairs of particles in the nucleus.

In chap. III one aspect of these intranuclear forces has already been discussed. There the phenomena of scattering and disintegration have been treated graphically by use of the "potential barrier". This curve (p. 48) gives the potential energy of an α particle, proton, or other particle, in terms of its position when in the neighbourhood of a nucleus. In the finer analysis, in any given position, such a particle will possess energy (positive or negative) by virtue of its proximity to each of the nuclear constituents separately regarded, whilst, if these constituents be in motion, periodic variations in the energy of the particle should be taken into account. In the potential barrier, however, only the time average of the

[a] Bartlett, *Nature*, 130, 165, 1932; *Phys. Rev.* 42, 145, 1932.
[b] Landé, *Phys. Rev.* 43, 620, 624, 1933.
[c] Elsasser, *J. Physique*, 4, 549, 1933.

total potential energy arising from individual actions upon the exterior particle is represented—and, for most purposes, this is sufficient elaboration. This general viewpoint may be extended to the nuclear particles themselves. Each of these is regarded[a] as possessing some potential energy because of its position near to all the others, taken one by one: the total negative energy of all the particles (reckoned from a zero corresponding to their complete separation, at rest) is the (positive) binding energy of the system. In order to calculate this quantity,[b] therefore, it is only necessary to know the force function for each of the pairs capable of being formed amongst the elementary units of structure—in the present case the force functions describing, respectively, the interactions between neutron and neutron, proton and proton and neutron and proton. Both the absolute magnitude of the force (or mutual potential energy, in terms of the above zero) at a specified separation and the form of its variation with increasing separation must be given by the appropriate function. At present, however, existing knowledge is not complete in either respect. At large distances the proton-proton interaction is known[c] to approximate very closely indeed to repulsion according to the inverse square (Coulomb) law, the force of repulsion at separation r being e^2/r^2, but direct confirmation of this law for small distances is still lacking. Otherwise, the best attested conclusion concerns the relative magnitudes of the various potential energies at a vaguely determined distance, the mean separation of the particles in the nucleus.[d] At this distance the negative potential energy corresponding to attractive forces must obviously be in excess of the positive potential energy

[a] This is not the only type of general assumption which may be made—but it is the simplest. See Peierls, *Reports on Progress in Physics*, 2, 27, 1936; also Bohr, *Nature*, 137, 344, 1936. Bohr's suggestions are further considered in § 38.

[b] For a summary concerning this aspect of the problem see v. Weizsäcker, *Phys. Z.* 36, 779, 1935, also Massey and Mohr, *Proc. Roy. Soc.* 152, 693, 1935; *Nature*, 136, 141, 1935; Thomas, *Phys. Rev.* 47, 903, 1935.

[c] Gerthsen, *Ann. Physik*, 9, 769, 1931.

[d] It is generally assumed that this distance does not vary much from one nucleus to another—nor, if there is any significance in further particularisation, from point to point in a heavy nucleus. See Gamow, *Structure of Atomic Nuclei and Nuclear Transformations*, 1936, chap. II, § 2.

characteristic of the inverse square law repulsion between protons and of any other repulsive forces operative. To determine whether the main portion of this negative potential energy is to be ascribed to the interaction of neutrons with neutrons, of neutrons with protons, or to any non-coulombian interaction between pairs of protons at intranuclear distances requires, however, further consideration. That it is to be ascribed chiefly to neutron-proton interaction is suggested by the fact that only one type of nucleus of mass number 2 is known, and that this nucleus, the deuteron, has charge number 1; it is the system neutron-proton, rather than either of the combinations neutron-neutron or proton-proton, which exhibits stability in nature. A more detailed examination[a] of the mass and charge numbers of the heavier nuclei confirms this broad conclusion. Likewise, on evidence from exact masses, it is now generally assumed that at the mean separation of nuclear particles the positive potential energy due to the repulsion between protons is relatively insignificant. However, the mutual potential energy of pairs of neutrons and that due to non-coulombian attraction between protons are clearly both considerable.[b]

Having established an order of importance amongst the various energies of interaction the next step consists in determining the "law of force" characteristic of the most important of them; in this case in fixing the law of force between neutron and proton. Already, many suggestions[c] have been made, all of them essentially empirical in nature. That this is necessarily so is obvious if neutron and proton are each regarded as elementary particles—we are then concerned with a type of interaction concerning which there is no previous information: it is almost equally obvious on the older view (variously maintained for a time, but now generally

[a] Gamow, *loc. cit.*

[b] See, Feenberg, *Phys. Rev.* 47, 850, 1935; Young, *ibid.* 47, 972, 1935; Feenberg and Knipp, *ibid.* 48, 906, 1935; Young, *ibid.* 48, 913, 1935; Present, *ibid.* 49, 640, 1936; Massey and Mohr, *Proc. Roy. Soc.* 156, 634, 1936.

[c] Wigner, *Phys. Rev.* 43, 252, 1933; Heisenberg, *Z. Physik*, 77, 1; *ibid.* 78, 156, 1932; *ibid.* 80, 587, 1933; Majorana, *ibid.* 82, 137, 1933; Breit and Wigner, *Phys. Rev.* 48, 918, 1935.

abandoned—cf. p. 133) of the neutron as a close combination of proton and electron (cf. p. 61). Admittedly empirical, however, the law of force suggested by Heisenberg, and subsequently modified by Majorana, is based upon an analogy which in many respects is particularly close. This analogy compares the deuteron, the stable combination of neutron and proton, with the positive ion of molecular hydrogen which may be regarded as a similar combination of proton and neutral hydrogen atom. The quantum dynamics of the latter structure is now well established.[a] It may be discussed, in terms of the potential barrier for electrons, roughly as follows. Three particles are considered, two hydrogen nuclei at a fixed (arbitrary) distance apart and a single electron, "somewhere" in their immediate neighbourhood. Between the two nuclei there will "exist" a potential barrier in respect of this electron of finite height and breadth. On this account, even if it be given, initially, that the electron is near one nucleus rather than the other, there will be a definite probability, per unit time, of transition, whereby the electron passes over to the other nucleus: the potential barrier has finite transparency. There is, in fact, a characteristic frequency of exchange for the electron, which may be evaluated as a function of the distance between the nuclei. Suppose that the exchange frequency is $\omega(r)$ when this distance is r. Then exchange energy, in amount $h\omega(r)$, is associated with the system because of this effect and an attractive force, $h\omega'(r)$, arises in consequence. We may regard this force as responsible for the stability of the system—which would not otherwise hold together. Concerning the molecule in its ground state it is meaningless to ask with which nucleus the electron is at any instant associated; it must be regarded, without further qualification, as completely shared between the two. If the introduction of the idea of exchange forces has resulted in a satisfactory theory of this particular ion (the simplest molecular unit) it has also furnished the clue to the explanation of a general fact of molecular chemistry, the so-called

[a] Cf. Mott, *An Outline of Wave Mechanics*, chap. VI, § 1, 1930. Richardson, *Proc. Roy. Soc.* 152, 503, 1935.

saturation of valency bonds.[a] In each respect[b] it has its application, by analogy, in the nuclear domain. In Majorana's variant of the theory saturation occurs when every neutron is bound to two protons, and vice versa. In this result the great stability of the α particle and nuclear masses closely integral on the oxygen scale (p. 60) appear to receive a natural explanation.

So far the only evidence in favour of the Heisenberg-Majorana theory is the degree of correlation amongst the ascertained facts of nuclear physics which may be exhibited by its aid. One enormous advantage of a force function of the exchange type is that it simulates a force of repulsion at small distances; no additional hypothesis is necessary to ensure stability for the system against the possibility of self-annihilation by condensation. In the original form of the theory—as already indicated—actual exchange of charge between neutron and proton was not seriously considered as the physical basis of forces of this type;[c] there was in existence a moderately simple force function with a precise significance in another connection, and it was enough that a function of this mathematical form should have considerable success in a new domain. Now, however, the situation is not quite so simple as this description would suggest. Fermi's theory of β disintegration (§ 28) does assume a real transition from proton to neutron (or vice versa) in every act of ejection of a (positive or negative) electron from a radioactive nucleus. It was clear at the outset—if the validity of neither theory were questioned (see p. 137)—that the two theories, of nuclear constitution and of β disintegration, should have a common starting-point in the law of force which is under

[a] London, *Z. Physik*, 46, 455; *ibid.* 50, 24, 1928; cf. Born, *Atomic Physics*, p. 251, 1935.

[b] By contrast it is interesting to compare the methods employed in discussions of the stability of the H_3^+ ion, on the one hand (Coulson, *Proc. Camb. Phil. Soc.* 31, 244, 1935), and of the nucleus, H^3, on the other (Massey and Mohr, *Proc. Roy. Soc.* 152, 693, 1935).

[c] Massey and Mohr (*Proc. Roy. Soc.* 148, 206, 1935) have concluded that this assumption, of actual exchange of charge associated with a small transfer of mass, is definitely untenable when confronted with experimental data concerning the photo-disintegration of the deuteron (§ 44).

discussion. At first, numerical considerations appeared to indicate that the theories in question failed to pass the test of this comparison,[a] but, recently, modifications have been suggested[b] in an effort to remove the inconsistencies thus revealed. It is perhaps too early, as yet, to judge fully of their success. From another point of view the attempt is being made to base the assumption of "exchange" forces on something more than mere analogy, and so provide a coherent theory of neutron and proton and their mutual actions.[c]

Experimental evidence for or against a force function of the exchange type might be expected to be provided by data concerning the angular distributions characteristic of the elastic scattering of neutrons by hydrogen nuclei, that is by the angular distribution determined for scattered neutrons or for the protons projected in such encounters. The relevance of similar investigations in respect of α particle scattering has already been discussed (p. 36). At the present time, however, the data in question are neither sufficiently extensive nor precise to warrant detailed consideration in this connection. The most satisfactory method of obtaining information is by means of the expansion chamber, through observations of the directions of projection of protons in a hydrogenous gas. Essentially tedious, when statistical results are required, this method, as applied to neutron-proton collisions, has until now suffered from two further disabilities, the weakness of the neutron sources regularly available and the wide spread of velocity amongst the neutrons which these sources provide. When more intense sources of mono-energetic neutrons can be employed (see p. 172) it is likely that very significant conclusions will be reached. The most complete results as yet obtained[d] appear to indicate a distribution of protons which is not isotropic when referred to a frame of reference in which the centre of gravity of the colliding particles is at rest. In so far as this conclusion can be relied upon it is at variance

[a] Tamm, Iwanenko, *Nature*, 133, 981, 1934.

[b] Bethe, *Int. Conf. Phys.* 1, 66, 1935; Konopinski and Uhlenbeck, *Phys. Rev.* 48, 7, 107, 1935.

[c] Wataghin, *Phys. Rev.* 48, 284, 1935.

[d] Harkins, Gans, Kamen and Newson, *Phys. Rev.* 47, 511, 1935.

with certain assumptions which are commonly made in theoretical discussions—or, it may be, more fundamentally at variance with the theory itself. It has generally been assumed that the range of strong interaction between neutron and proton is small compared with the radius of the deuteron;[a] then, on this assumption, it is calculated[b] that any asymmetry in the angular distribution of projected protons should be below the range of experimental detection, unless neutrons of energy greater than 2×10^7 electron volts be employed. The mean energy of the neutrons used for the expansion chamber experiments of Harkins, Gans, Kamen and Newson was of the order of 6×10^6 electron volts. Obviously, further data are urgently required.

Data concerning the angular distributions of the neutrons and protons produced in the photo-disintegration of deuterium (chap. XIII) are at present very meagre,[c] but here, also, considerable extension may be expected in the near future. The distributions in this case depend upon the quantum specification of the ground state of the deuteron; in the first analysis they may lead to a decision whether or not this state has zero orbital momentum, as simple theories would predict. At a later stage present assumptions concerning the law of force between the particles may need to be modified so as to lead to a specification of the ground state consistent with the extended observations upon this phenomenon.

Direct experimental evidence regarding a possible non-coulombian term in the law of force between protons must necessarily come from a study of the scattering of protons in hydrogen. A beginning in this respect has been made by White[d] using artificially accelerated particles. The indications of anomalous scattering which he obtained have recently been discussed, in terms of a simple attractive interaction, by

[a] For the resolution of this apparent paradox, see Peierls, *Reports on Progress in Physics*, 2, 27 (§ 5), 1936.

[b] See, for example, Bethe and Peierls, *Proc. Roy. Soc.* 149, 176, 1935.

[c] Chadwick and Goldhaber, *Proc. Roy. Soc.* 151, 479, 1935; Chadwick, Feather and Bretscher, *ibid.* in course of publication.

[d] White, *Phys. Rev.* 47, 573, 1935; *ibid.* 49, 309, 1936; see also Wells, *ibid.* 47, 591, 1935; Tuve, Heydenburg and Hafstad, *ibid.* 49, 402, 1936.

Present.[a] Practically all conclusions in this field must depend upon future experiment.

Finally, it may be said that future experiment is very unlikely to furnish direct information concerning the third fundamental force function necessary for the exact solution of nuclear problems—the force function describing the possible interaction between neutron and neutron. All evidence for this must clearly be of a more indirect nature.

[a] Present, *Phys. Rev.* 48, 919, 1935; *ibid.* 49, 201, 1936; see also Breit and Condon, *Phys. Rev.* 49, 866, 1936.

PART TWO

CHAPTER V

NUCLEAR CHARGE AND MASS

§ 14. *Mass and charge numbers.* Part II of this book deals particularly with the stable nuclei: the present section is intended to form an introduction to that subject by discussing limits of stability empirically, in terms of nuclear mass and charge numbers as already defined. For this purpose it is necessary that these constants (A and Z) be known both for those nuclear species believed to be stable and also for the species for which instability has been proved. During the last two years, since the initial discovery of the phenomenon of artificial radioactivity (see § 26), a very great deal of new material has become available for such a survey; without it any conclusions which might be reached would appear to be much less securely grounded.

From the point of view of physics the determination of nuclear charge in any case depends upon the analysis of the spectrum of characteristic X-radiation (p. 22) or upon the determination of the absolute amount of α particle scattering under determined conditions (p. 35). With many substances neither of these experiments is possible: then—and very frequently in other cases also—purely chemical methods are employed. No ambiguity is introduced by this procedure (except in particular circumstances—as, for example, with the rare earth elements) since the identification of nuclear charge number with the ordinal number of the element in the periodic classification of Mendeléeff is so thoroughly established. As regards mass number, however, physical methods alone are of any use for the determination: in general the various isotopes of an element are chemically indistinguishable. When the methods of mass spectroscopy (p. 10) fail evidence in this direction may be obtained only from optical spectra (in suitable cases)[a] or from a knowledge of the nuclear transformation involved (§ 16), if the species in question is produced by such a process.

The values of A and Z for the stable species are given in Table 1, which also includes a small number of doubtful[b] cases suitably distinguished "(?)". It will be noticed that, between 0 and 83, only one[c] value of Z ($Z=61$) is missing from this table (and this atomic number obviously belongs to the—stable or unstable—product of the β activity of neodymium, $Z=60$, if the latter activity be regarded as established).[d] Similarly, all values of the mass number from 1 to 209 are represented with the exception of the mass number 8. In itself this is strong evidence in favour of the assumption that very few stable species are capable of existence which are not included in the table: broadly speaking two facts are responsible for such doubt as remains on this point. In the first place the frequent occurrence of isobaric pairs and triads (p. 60) leaves open the possibility of the existence of isobars in other cases also, and, secondly, the large range of relative abundance[e] characteristic of those species which are known[f] provokes the suspicion that somewhat more sensitive means of detection might well extend that range by the discovery of nuclear species of still smaller abundance than any hitherto detected. Data concerning artificial radioactivity, on the other hand, provide a great deal of evidence which points to the opposite conclusion; the doubt which has been expressed above is considerably lessened on this account.

[a] Mulliken, *Nature*, 113, 423, 489, 1924; Giauque and Johnston, *ibid.* 123, 318, 831, 1929; Babcock, *ibid.* 123, 761, 1929; Naudé, *Phys. Rev.* 36, 333, 1930; King and Birge, *Nature*, 124, 127, 1929; Urey, Brickwedde and Murphy, *Phys. Rev.* 40, 1, 1932; Schüler and Jones, *Naturwiss.* 20, 171, 1932; Wehrli, *Helv. Phys. Acta*, 7, 611, 1934; Venkatesachar and Sibaiya, *Nature*, 136, 65, 437, 1935; Fuchs and Kopfermann, *Naturwiss.* 23, 372, 1935.

[b] The doubt concerns the stability of the particular nucleus, not its existence.

[c] $Z=43$ (masurium) is included, on the assumption that the existence of this chemical element is definitely established. It is, then, the one existing element of which nothing is known regarding its isotopic constitution.

[d] Libby, *Phys. Rev.* 46, 196, 1934.

[e] Aston, *Mass Spectra and Isotopes*, chap. XIV, 1933.

[f] The ratio of the numbers of nuclei O^{16} and Xe^{124} upon earth has been given as about $10^{17}:1$.

Table 1

Symbol	Z	A	Symbol	Z	A
n	0	1 (?)	Cd	48	106, 108, 110, 111,
H	1	1, 2, 3			112, 113, 114, 116
He	2	3, 4	In	49	113, 115
Li	3	5 (?), 6, 7	Sn	50	112, 114, 115, 116,
Be	4	9, 10(?)			117, 118, 119, 120,
B	5	10, 11			122, 124
C	6	12, 13	Sb	51	121, 123
N	7	14, 15	Te	52	120, 122, 123, 124,
O	8	16, 17, 18			125, 126, 128, 130
F	9	19	I	53	127
Ne	10	20, 21, 22	Xe	54	124, 126, 128, 129,
Na	11	22 (?), 23			130, 131, 132, 134,
Mg	12	24, 25, 26			136
Al	13	27	Cs	55	133
Si	14	28, 29, 30	Ba	56	130, 132, 134, 135,
P	15	31			136, 137, 138
S	16	32, 33, 34	La	57	139
Cl	17	35, 37	Ce	58	136, 138, 140, 142
A	18	36, 38, 40	Pr	59	141
K	19	39, 41	Nd	60	142, 143, 144, 145,
Ca	20	40, 42, 43, 44			146, 148, 150
Sc	21	45	Sm	62	144, 147, 148, 149,
Ti	22	46, 47, 48, 49, 50			150, 152, 154
V	23	51	Eu	63	151, 153
Cr	24	50, 52, 53, 54	Gd	64	155, 156, 157, 158,
Mn	25	55			160
Fe	26	54, 56, 57, 58	Tb	65	159
Co	27	57, 59	Dy	66	161, 162, 163, 164
Ni	28	58, 60, 61, 62, 64	Ho	67	165
Cu	29	63, 65	Er	68	166, 167, 168, 170
Zn	30	64, 66, 67, 68, 70	Tm	69	169
Ga	31	69, 71	Yb	70	171, 172, 173, 174,
Ge	32	70, 72, 73, 74, 76			176
As	33	75	Lu	71	175
Se	34	74, 76, 77, 78, 80, 82	Hf	72	176, 177, 178, 179,
Br	35	79, 81			180
Kr	36	78, 80, 82, 83, 84,	Ta	73	181
		86	W	74	182, 183, 184, 186
Rb	37	85, 87	Re	75	185, 187
Sr	38	84, 86, 87, 88	Os	76	186, 187, 188, 189,
Y	39	89			190, 192
Zr	40	90, 91, 92, 94, 96	Ir	77	191, 193
Nb	41	93	Pt	78	192, 194, 195, 196,
Mo	42	92, 94, 95, 96, 97,			198
		98, 100, 102	Au	79	197
Ma	43	—	Hg	80	196, 197, 198, 199,
Ru	44	96, 98, 99, 100,			200, 201, 202, 203,
		101, 102, 104			204
Rh	45	103	Tl	81	203, 205
Pd	46	102, 104, 105, 106,	Pb	82	204, 206, 207, 208
		108, 110			
Ag	47	107, 109	Bi	83	209

Before discussing this evidence further, however, it will be well to return to the consideration of the radioelements of high atomic weight—and of the small additional number of unstable elements of long life of which the activity has been known for a considerable time. Table 2 contains the relevant information. The charge numbers which are involved are 19, 37, 60, 62, together with all the numbers from 81 to 92, inclusive, with the exception of 85 and 87.[a] Whilst, therefore,

Table 2

Symbol	Z	A	Symbol	Z	A
K	19	40	Rn	86	219, 220, 222
Rb	37	86(?)	Ra	88	223, 224, 226, 228
Nd	60	147(?)	Ac	89	227, 228
Sm	62	151(?)	Th	90	227, 228, 230, 231, 232, 234
Tl	81	207, 208, 210			
Pb	82	210, 211, 212, 214	Pa	91	231, 234
Bi	83	210, 211, 212, 214	U	92	234, 235, 238
Po	84	210, 211, 212, 214, 215, 216, 218			

the elements thallium (81), lead (82) and bismuth (83) include both stable and unstable species, for atomic numbers greater than 83, instability appears to be the universal rule. That our knowledge of this instability in the phenomena of "natural" radioactivity should be limited to species with $Z \leqslant 92$ may be regarded as something of an accident: in fact it appears likely that unstable trans-uranic elements have now been produced "artificially" by neutron bombardment (p. 129). At any rate there is no evidence for a further range of values of Z greater than 92 for which stable nuclei may again become possible.

The evidence from the newly discovered radioactive species of short life ("artificial radioelements") in like manner concerns the limitation of stability, for $Z \leqslant 80$, to those species having particular values of A. At the present time, for the moderately heavy elements, at least, this evidence is the more conclusive when the atomic number, Z, is odd, than

[a] According to Gueben (*Ann. Soc. Sci. Bruxelles*, 53, 115, 1933) the species $Z=87$, $A=224$ is formed in a very rare α mode of disintegration of mesothorium 2.

when it is even. This is the natural result of the fact, long recognised, that the phenomenon of isotopy is much more pronounced amongst elements of even atomic number than with those of odd: in the latter case many elements are simple and none heavier than lithium possess more than two stable isotopes. For odd numbered elements heavier than phosphorus ($Z=15$), when two stable isotopes occur, invariably mass numbers also are odd and differ by two units. Now the data in question show that unstable species of odd atomic number are frequently produced, not only having mass numbers greater and less than the mass numbers of the heavier and lighter of the isotopic stable species, respectively —as the unstable chlorine isotopes, Cl^{34} and Cl^{38}, stand in reference to Cl^{35} and Cl^{37}—but also with the "missing" intermediate mass numbers as well. The recently discovered unstable species Cu^{64}, Ga^{70}, Br^{80}, Ag^{108} and In^{114} provide examples of this result: for this reason it is natural to conclude that with odd numbered, two-isotope, elements in general the "missing" species has never been found precisely for the reason that it is definitely unstable. In one case, only, might this rule appear to admit of exception: the accurate investigations of Nier[a] and of Brewer[b] make it reasonably certain that the isotope K^{40} represents about 1 part in 8500 of the potassium upon earth. On the other hand the view has been gaining ground for some time[c] that the natural radioactivity of potassium must be ascribed to this species. In Table 2 a like hypothesis is made—though without similar experimental confirmation—regarding the origin of the natural activity of rubidium. The rule, in fact, has not been broken.

For the lighter elements the evidence from artificial radioactivity is more complete; it is given in the usual manner in Table 3 and in Fig. 5 is exhibited with increased forcefulness. Here known nuclei, stable and unstable,[d] are represented on a two-dimensional diagram in which axes of Z and $A-Z$ are

[a] Nier, *Phys. Rev.* 48, 283, 1935. [b] Brewer, *Phys. Rev.* 48, 640, 1935.

[c] Hevesy, *Nature,* 135, 96, 1935; Klemperer, *Proc. Roy. Soc.* 148, 638, 1935.

[d] As previously mentioned, certain doubtful cases still exist; attention is drawn to these in the tables, the corresponding points are included in Fig. 5 without further differentiation.

employed. (On current views the number of protons in the nucleus is thus plotted against the number of neutrons.) Stable nuclei are indicated by full circles, unstable nuclei by open circles. It will be observed that the former lie in a well defined region, flanked on either side by the open circles.

Table 3

Symbol	Z	A	Symbol	Z	A	Symbol	Z	A
Li	3	8	O	8	15, 19	Al	13	26, 28, 29
Be	4	8(?)	F	9	17, 20	Si	14	27, 31
B	5	12	Ne	10	23	P	15	30, 32
C	6	11, 14	Na	11	22(?), 24	S	16	31(?), 35
N	7	13, 16	Mg	12	27	Cl	17	34, 38

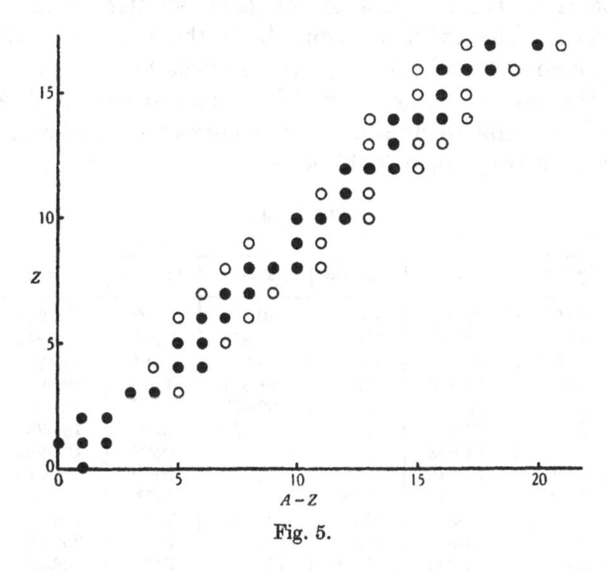

Fig. 5.

Since only integral values of the variables are physically significant, it is obvious that the possibility of discovery of further stable species in this range is very restricted indeed. Again, it is natural to conclude on this evidence, and on the partial evidence at present available in the range of the heavier elements, that eventually a figure such as the above will be constructed which will include all the elements—and exhibit the stable species at present known as the great majority of those possible in the actual order of things.

§ 15. *Nuclear mass: the mass spectrograph.* Until quite recently experiments with the mass spectrograph provided the only method of determining the "exact" masses of nuclei—as also of fixing the corresponding mass numbers. Within the last few years, however, the enormous increase in the number of nuclear transformations accurately investigated has provided a check—and, in fact, an independent means of determination (if one mass ratio be given)—which has proved more powerful still, whenever it could properly be applied. Also, as previously stated, a certain amount of evidence has arisen from the analysis of the band spectra of diatomic molecules. The evidence from both these sources will be considered in the next section; its effect here is merely to limit the data which are quoted from mass spectrum observations in general to nuclear species heavier than aluminium. These data are given in Table 4; they comprise the determined

Table 4

Species	Mass	Species	Mass	Species	Mass
Si^{28}	27·982	Br^{81}	80·930	Xe^{134}	133·929
P^{31}	30·983	Kr^{78}	77·926	Cs^{133}	132·933
Cl^{35}	34·983	Kr^{80}	79·926	Ba^{138}	137·916
Cl^{37}	36·980	Kr^{82}	81·927	Ta^{181}	180·928
A^{36}	35·976	Kr^{83}	82·927	Re^{187}	186·981
A^{40}	39·971	Kr^{84}	83·928	Os^{190}	189·981
Cr^{52}	51·948	Kr^{86}	85·929	Os^{192}	191·981
Ni^{58}	57·942	Nb^{93}	92·926	Hg^{200}	200·016
Ni^{60}	59·940	Mo^{98}	97·946	Tl^{203}	203·037
Zn^{64}	63·937	Mo^{100}	99·945	Tl^{205}	205·037
As^{75}	74·934	Sn^{120}	119·912	Pb^{206}	206·00
Se^{78}	77·938	Te^{126}	125·937	Pb^{208}	208·00
Se^{80}	79·942	Te^{128}	127·936	Th^{232}	232·070
Br^{79}	78·929	I^{127}	126·932	U^{238}	238·088

masses, on the basis of $O^{16} = 16$, of the neutral atoms of the various species which have been sufficiently carefully investigated. The procedure of quoting atomic rather than nuclear masses in a table such as this is standard practice and has been strongly advocated.[a] To the writer it appears slightly illogical when, as in the present instance, the masses are to be employed entirely in considerations of nuclear structure and transformations: to quote nuclear masses and retain, if need be, the integral standard $O^{16} = 16$ for the neutral atom (as a gesture of respect for the days of stoicheiometrical innocence!) would appear entirely preferable. However, if any (energy) equation in nuclear masses is correct, the corresponding equation in which atomic masses occur will be similarly valid,[b] but it does appear slightly illogical to employ these masses in calculations relative to purely nuclear interactions when knowledge is completely lacking (and entirely irrelevant) as regards incidental and wholly temporary changes in extranuclear configuration. If nuclear masses be required, the quantity $5{\cdot}47 . 10^{-4}\, Z$ should be subtracted from each of the masses in Table 4.

Recently the beginnings have been made with a new and more accurate determination of the relative masses of the light elements by mass spectrograph methods. Table 5 includes the provisional values given by Aston[c] as a result of

Table 5

Species	Mass	Species	Mass
H^1	$1{\cdot}0081_2$	F^{19}	19·0045
H^2	$2{\cdot}0147_1$	Ne^{20}	19·9986
He^4	4·0039	Al^{27}	(26·9909)
B^{10}	10·0161	Si^{28}	(27·9860)
C^{12}	12·0035	Si^{29}	28·9864
N^{14}	14·0073	A^{40}	39·9754
O^{16}	16·0000		

this work. Already the general agreement with the data from disintegration experiments has been materially improved; how far that improvement will be carried is a matter for future experiment to decide.

[a] *Solvay Conference Report*, 1934.

[b] Except in so far as the equation represents a nuclear change in which electrons are produced. Then, although the true ("nuclear") mass of electron and positron is in each case 0·000547, as may easily be seen the effective ("atomic") mass of the latter must be taken as 0·001094 and that of the former as zero.

[c] Aston, *Nature*, 135, 541, 1935; *ibid.* 137, 357, 613, 1936.

§ 16. *Evidence from transformation experiments and from other sources.* The first serious difficulty in the way of accepting the values of nuclear mass deduced by the earlier methods arose in the case of beryllium. For this nucleus, Be^9, Bainbridge[a] obtained a mass greater than the sum of the masses of two α particles and a neutron, according to the values then accepted. On the other hand, careful experiments[b] failed to reveal the natural activity which was to be expected (cf. p. 59) on the basis of this result. Gradually other discrepancies were detected and independent attempts were made by Oliphant, Kempton and Rutherford[c] and by Bethe[d]—from slightly different points of view—to remove them by the proposal of a set of masses based primarily upon disintegration data. The masses given in Table 6 have been chosen in view of this work, and of later discussions of a similar nature.[e] One result of their acceptance is that Be^9 comes to be regarded as a normally stable species.

Table 6

Species	Mass	Species	Mass	Species	Mass
n^1	1·0090	B^{10}	10·0161	Ne^{20}	19·9986
H^1	1·0081	B^{11}	11·0128	Ne^{21}	21·0007
H^2	2·0147	C^{12}	12·0035	Ne^{22}	21·9987
H^3	3·0170	C^{13}	13·0073	Na^{23}	22·9971
He^3	3·0170	N^{14}	14·0073	Mg^{24}	23·9938
He^4	4·0039	N^{15}	15·0048	Mg^{25}	24·9946
Li^6	6·0168	O^{16}	16·0000	Mg^{26}	25·9909
Li^7	7·0179	O^{17}	17·0046	Al^{27}	26·9909
Be^9	9·0149	O^{18}	?	Si^{30}	29·9844
Be^{10}	10·0164	F^{19}	19·0045	P^{31}	30·9844

At this stage it is worth while discussing more fully the method of employing the observations recorded in disintegration experiments in calculations such as have been described.

[a] Bainbridge, *Phys. Rev.* 43, 367, 1933.

[b] Rayleigh, *Nature*, 131, 724, 1933; Evans and Henderson, *Phys. Rev.* 44, 59, 1933; Gans, Harkins and Newson, *ibid.* 44, 310, 1933.

[c] Oliphant, Kempton and Rutherford, *Proc. Roy. Soc.* 150, 241, 1935.

[d] Bethe, *Phys. Rev.* 47, 633, 1935.

[e] Cf. Bonner and Brubaker, *Phys. Rev.* 49, 19, 1936; Oliphant, *Nature*, 137, 396, 1936; Cockcroft and Lewis, *Proc. Roy. Soc.* 154, 246, 261, 1936. (An independent evaluation of some of the masses here quoted has been made by Pollard and Brasefield, *Nature*, 137, 943, 1936.)

Certain assumptions are obviously involved. It is assumed, for instance, that the nature of the disintegration is known—and it is postulated that the mechanical conservation laws, of momentum and of mass-energy (p. 29), apply. Here we shall consider only the simplest type of disintegration which may be represented by the nuclear scheme

$$_eX^a + _fx^b \rightarrow _gY^c + _hy^d \qquad \text{......(14);}$$

it involves capture of the incident particle x by the nucleus X and results in the ejection of the particle y, leaving Y as the heavier product nucleus or residue. Superscripts a, b, c, d indicate the mass numbers and subscripts e, f, g, h the charge numbers[a] of the nuclei. Conservation of charge and of the number of heavy nuclear particles is expressed in the equations

$$a+b=c+d; \quad e+f=g+h \qquad \text{......(15).}$$

A scheme such as (14) merely indicates the course of the transformation; however, the same symbols may be used in a nuclear equation to express quantitatively the amounts of energy involved in the process. In the equation

$$_eX^a + _fx^b = _gY^c + _hy^d + Q \qquad \text{......(16)}$$

the symbols represent the exact masses of the nuclei—and Q the mass equivalent of the energy released in the transformation. If E_x, E_Y and E_y are the kinetic energies of the particles x, Y and y, in any particular instance of the transformation, then (the initial energy of the nucleus X being effectively zero)

$$Q = E_Y + E_y - E_x \qquad \text{......(17).}$$

Any experimental investigation must necessarily begin by establishing the result that the same value of Q is given by (17) in respect of all observed instances of the transformation in question. In general this result is not established without qualification: Q values for different cases of one and the same transformation usually belong to a small number of distinct energies Q_1, Q_2, ... As will appear later (§ 33), all but the greatest of these (Q_1) are interpreted in terms of temporary

[a] Ideally, charge numbers are redundant in any expression in which chemical symbols are employed, being uniquely determined by the chemical nature of a substance. Usually, however, they are included, if only for sake of convenience.

excitation of the product nucleus Y; for mass calculations, however, only Q_1 (the energy released in the formation of an unexcited residue) is of primary[a] significance. In establishing (17) the method of experimenting is as follows. For a given value of E_x, E_y is determined (p. 83) in its dependence upon θ, the angle between the directions of incidence of x and of projection of y. The validity of the conservation laws and the assumption of a single value of Q lead to the relation

$$(c+d)\,E_y - 2\,(bdE_yE_x)^{\frac{1}{2}}\cos\theta - (c-b)\,E_x = cQ \quad \ldots(18),$$

in which initial mass numbers, with sufficient accuracy, are employed in place of the exact masses of the nuclei involved. Bothe[b] first showed, in respect of the disintegration of boron by α particles, that the angular variation of kinetic energy, for the disintegration particles corresponding to each particular value of Q, was of the predicted type. More recently abundant confirmation of this result has been obtained in many cases:[c] it is now universally accepted in calculating disintegration energies for the different reactions. Equation (18) also predicts the type of variation of E_y with E_x for a given value of θ. Observations made in the forward direction ($\theta=0$), using bombarding α particles of different energies, were discussed from this point of view by Bothe[d] in 1928. Soon afterwards a number of cases of disintegration under α particle bombardment was examined in a similar manner with verification of (18). The first extensive experiments of this kind in which protons were used for the bombardment were made by Henderson, Livingston and Lawrence[e] in 1934. Again, the assumptions underlying (18) were found to be justified. Altogether, the method of determining mass differences from disintegration data by means of the relation (17) will be seen to be securely grounded, the only remaining

[a] Obviously the differences (Q_1-Q_2), (Q_1-Q_3) ... represent energies of excitation of the nucleus Y and a knowledge of them is of importance as a check on the assumed binding energy of Y. In general it cannot be imagined that the binding energy should be less than the greatest of these energy differences.

[b] Bothe, *Z. Physik*, 63, 381, 1930.

[c] See, for example, Pose, *Phys. Z.* 31, 943, 1930.

[d] Bothe, *Z. Physik*, 51, 613, 1928.

[e] Henderson, Livingston and Lawrence, *Phys. Rev.* 46, 38, 1934.

consideration concerns the methods of energy determination which in practice are possible. Here the matter of intensity introduces the gravest problem. Quite frequently the energy of the incident particles may be determined with accuracy—by standard methods of magnetic spectroscopy—but for the much less numerous particles which result from a disintegration this is never the case. Energies must be deduced from the ranges of these particles in air or some other more suitable medium. This necessarily introduces the possibility of error, since the requisite data connecting range and energy are often much less extensive and trustworthy than might be desired. Moreover, in the case of individual disintegration events investigated by means of the expansion chamber (as the energetics of almost all neutron-produced disintegrations have so far been investigated) there is no possibility of applying deflection methods even to the incident particles[a]—all energies must be deduced from determinations of range. Gradually this process, too, is losing some of its uncertainties: experimentally determined range-energy relations, extrapolated with the aid of more or less approximate theory, have been modified so as to yield self-consistent results when applied to chosen cases of nuclear disintegration;[b] in this form they may be employed in other cases with correspondingly greater confidence. As a result, whilst in most cases 10^6 electron volts would be considered a very large error in an accepted energy of disintegration, 0·001 mass units, its approximate equivalent, frequently lies well within the limits of uncertainty in the direct determination of nuclear mass.

The evidence from band spectra concerns the ratios of the masses of isotopic nuclei.[c] In brief the principles involved—and the broad assumptions which are made—are as follows. It is supposed that the general fields of force in two diatomic

[a] When these are neutrons, apart from the disintegration effects produced, there is no other feature from which the initial energies of effective particles may be estimated.

[b] Cf. Duncanson, *Proc. Camb. Phil. Soc.* 30, 102, 1934.

[c] Casimir (*Physica*, 1, 1073, 1934) has pointed out that the mass ratio calculated from these data is the ratio of the masses of the atomic ions formed by the nuclei together with all the electrons except those involved in the molecular binding.

molecules $X^a Y^p$ and $X^a Y^q$, which differ only in the nuclear mass of one of the constituent atoms, are identical. On this assumption the energies of corresponding nuclear rotation states (that is the energies of states having the same quantum specification for the two molecules) are inversely as the appropriate moments of inertia of the molecules, and energies of corresponding vibration states inversely as the square roots of the reduced masses ($ap/(a+p)$ and $aq/(a+q)$, respectively). In other words, writing ρ^2 for the reduced-mass ratio, corresponding frequencies of pure rotation transitions stand in the ratio $1 : \rho^2$ and corresponding frequencies of pure vibration transitions in the ratio $1 : \rho$.[a] Finally, the frequencies of electron transitions in the molecule are very nearly insensitive to the change in nuclear mass which is under consideration. In principle, therefore, since the only significant parameter which may not be obtained directly from the analysis of the spectra of molecules of a single type is the ratio ρ, the ratio of the reduced masses may be deduced from measurements of the displacements of corresponding lines in the band structure observed when molecules of both types are present—so long as these lines may be resolved and, when resolved, may be recognised in their true relation one to another. All this has successfully been achieved in a number of cases. In respect of the (infra-red) vibration-rotation spectra of H^1Cl and H^2Cl the analysis and calculations were carried out first of all by Hardy, Barker and Dennison;[b] in respect of several (optical) spectra of electronic bands, similar calculations have resulted in approximate values for the mass ratios $O^{16} : O^{17} : O^{18}$, $N^{14} : N^{15}$ and $C^{12} : C^{13}$.[c] These ratios have not so far been determined by the more direct methods of mass spectroscopy.[d]

In the finer details, however, the band spectrum method leaves much to be desired. The general relations above derived need modification to the extent of a number of small

[a] Cf. Born, *Atomic Physics*, pp. 236–46, 1935; Jevons, *Report on Band Spectra of Diatomic Molecules*, chap. X, 1932.

[b] Hardy, Barker and Dennison, *Phys. Rev.* 42, 279, 1932.

[c] For references, see p. 73.

[d] See, however, Bainbridge and Jordan, *Phys. Rev.* 49, 833; 50, 98, 1936.

correcting terms the importance of which had not until recently been fully investigated. Now, in so far as concerns the H^2/H^1 ratio, at least, the question has been thoroughly reviewed by Watson:[a] it is interesting to remark that the numerical value finally obtained is stated to lend weight to Aston's re-determination of the same ratio (p. 79).

The present chapter would be incomplete without some reference to atomic spectra in relation to mass determination. It will be recalled that the ratio of the masses of proton and electron was deduced[b] from the small difference in the series constants appropriate to the Balmer and Pickering series—once these spectra had been correctly attributed to hydrogen and ionised helium, respectively.[c] The result is involved that in the simplest (one-electron) atomic spectra, also, characteristic frequencies are determined by the reduced mass of the system. In all cases this mass is very close indeed to the rest mass of the electron; it is sufficiently different for a few of the lightest atoms, however, to be set in evidence directly by determinations of wave-length. The first proof of the concentration of heavy hydrogen (H^2) by Urey, Brickwedde and Murphy[d] was obtained in this way—in the spectrum of the enriched gas a second Balmer series appeared slightly shifted with respect to the first.

As above indicated, the problem of an isotope effect in atomic spectra is no longer a simple one when more than one extranuclear electron is present in the emitting system. For two and three electron systems the question has been discussed theoretically by Hughes and Eckart:[e] their general conclusion appears to be that the displacements observed[f] in the atomic spectra of neutral and once ionised lithium are in good agreement with theory—they are to be attributed purely to the difference in mass between the nuclei of the lithium isotopes. An isotope effect in the line spectra of

[a] Watson, *Phys. Rev.* 49, 70, 1936.
[b] Paschen, *Ann. Physik*, 50, 901, 1916.
[c] Bohr, *Nature*, 92, 231, 1913.
[d] Urey, Brickwedde and Murphy, *Phys. Rev.* 40, 1, 1932.
[e] Hughes and Eckart, *Phys. Rev.* 36, 694, 1930.
[f] Hughes, *Phys. Rev.* 38, 857, 1931.

heavy atoms was first established in the case of thallium by Schüler and Keyston[a] in 1931, since when many other examples of this effect have been recognised. Consideration of them belongs naturally to the next chapter; they may be regarded phenomenologically as a disturbing complication of the magnetic hyperfine structure pattern (§ 18) when the spectrum is that of a mixed element.[b] However, as will appear (p. 98), the origin of the effect is quite other than magnetic. From the standpoint of § 14 the fact that it may occur has proved fortunate in one respect: it has permitted the discovery of atomic species for which other evidence was, for the time being, lacking.[c] Isotopes of lead, platinum and iridium, discovered in this way, have since been confirmed by the standard methods of mass analysis.[d]

[a] Schüler and Keyston, *Z. Physik*, 70, 1, 1931.

[b] When "hyperfine" structure was first observed in the spectra of mercury and bismuth (Nagaoka, Sugiura and Mishima, *Jap. J. Phys.* 2, 121, 1923; *Nature*, 113, 459, 1924) the attempt was made to explain it as an isotope effect. The accepted explanation was given later by Pauli (p. 87).

[c] For references, see p. 73.

[d] The existence of In^{113} was first indicated by the band spectrum observations of Wehrli using indium iodide. This species, also, has since been detected with the mass spectrograph.

CHAPTER VI

ROTATIONAL ATTRIBUTES: MECHANICAL AND MAGNETIC MOMENTS OF NUCLEI

§ 17. *Historical.* The idea of electrons rotating in orbits around a massive nucleus was, as we have seen, essential to the dynamical atom model of Rutherford and Bohr; the experiments of Gerlach and Stern (p. 19) first showed that, for some atoms at least, this model had in addition to be furnished with a unique axis—of resultant magnetic moment and angular momentum. At the time (1922) there was no compelling reason to believe that these resultant moments need have any other components than those which standard theory would predict for individual electrons executing the orbital motions which were presupposed. The crude picture became that of electron orbits having fixed relative orientations within the atom. In 1924 the suggestion was made[a] that the nucleus itself ought also to be regarded as possessing a spin axis of a similar nature, and in consequence be restricted to certain definite orientations with respect to the rest of the atom. This suggestion, as will appear, was made in an attempt to explain the hyperfine structure of a number of lines in the optical spectra of mercury and bismuth. Essentially the same type of interpretation is current to-day. Soon afterwards a similar type of suggestion was advanced in respect of the electron.[b] Goudsmit and Uhlenbeck[c] proposed a physical interpretation of a successful formal scheme, which they had introduced in connection with multiplet structure and the Zeeman effect in atomic spectra, along these lines.[d]

[a] Pauli, *Naturwiss.* 12, 741, 1924.

[b] The suggestion of a magnetic or "spinning" electron is, in a sense, much older than is here suggested. Put forward on the basis of less and less experimental evidence, it may be traced through Compton and Rognley (*Phys. Rev.* 16, 464, 1920) back to Parson (*Smithsonian Misc. Coll.* 65, No. 11, 1915).

[c] Goudsmit and Uhlenbeck, *Naturwiss.* 13, 953, 1925; *Z. Physik*, 35, 618, 1926.

[d] Cf. Born, *Atomic Physics*, chap. VI, 1935.

Their interpretation, also, possesses an exact counterpart in the accepted theory of the effects in question. It is merely necessary to translate these suggestions of electron and nuclear spin into the language of modern theoretical physics in order to be in a position to discuss the experimental material with which the present chapter will deal.

In the first place, as regards electron spin, Dirac[a] has shown that to postulate "rotational" attributes for the electron does not, as might normally appear, involve the introduction of any additional hypothesis concerning the nature of the elementary particle: if this is to be describable by the methods of the wave mechanics in a way which admits the validity of the special (restricted) theory of relativity, then such attributes are essential to its nature. In the expression for the total energy of the particle a term occurs which is most naturally interpreted as due to an intrinsic magnetic moment $eh/4\pi mc$,[b] whilst the association of this with angular momentum $\frac{1}{2}h/2\pi$ may be demonstrated by standard methods. The fact that the magnetic moment of the proton is not similarly given by $eh/4\pi Mc$ (the protonic mass, M, replacing m, the mass of the electron, in the above expression), although the associated mechanical moment is the same ($\frac{1}{2}h/2\pi$), clearly indicates that, within the framework of Dirac's theory, at least, the proton cannot also be regarded as an elementary particle. For further consideration of this question, however, the reader is referred to other discussions,[c] here he must return to a more general consideration of atomic and nuclear moments from the new point of view.

From that point of view almost the whole emphasis moves to the energy of the system; for which experiment shows quantisation whenever the appeal to experiment can be directly made. The experiment of Stern and Gerlach, for example, provides evidence for the quantisation of magnetic

[a] Dirac, *Proc. Roy. Soc.* 117, 610; *ibid.* 118, 351, 1928; *Phys. Z.* 29, 561, 1928.

[b] A term which at first appeared to require the existence of an *electric* dipole moment has since been shown by Lees (*Proc. Camb. Phil. Soc.* 31, 94, 1935) to disappear in a more rigorous treatment.

[c] Cf. Gamow, *Structure of Atomic Nuclei and Nuclear Transformations*, 1936, chap. III, § 2.

potential energy, simply,[a] rather than for the precise orientation of atoms in an external field.[b] Similarly, from this point of view, the spatio-temporal atom model, also, loses much of its definiteness; in so far as it is retained, spherically symmetrical distributions of charge density replace electrons in circular orbits as the simplest configurations. There is one respect, however, important for present discussions, in which these two basic configurations are obviously not equivalent: spin moments may be associated with circular orbits, but not necessarily with a spherical distribution of charge. On the other hand, as already indicated, the spin of the electron now enters into the scheme, and resultant moments deduced from the Stern-Gerlach experiment are regarded as possessing components due to electron spin as well as possible nuclear components[c]—and components arising in "orbital" motion, when the distribution of charge is no longer spherically symmetrical and the flow of electron current is not everywhere zero.

Having made this brief attempt to exhibit the change in viewpoint of the theorist during the past twenty years, one further remark is in place. Although the orbital model of the atom is no longer capable of providing a satisfactory basis of explanation of all phenomena, equally clearly rotational properties of the atom need to be considered. For this purpose vector representation is able to give, in a simple geometrical fashion, all the results of importance. Amongst experimental physicists probably there will always be those who think of this method of representation to some extent in terms of an orbital model, or at least in connection with certain ideas of the precise orientation of rotating systems. If the phraseology of succeeding sections indicates this attitude in the writer it may not be difficult for those who so desire to interpret his statements more in accordance with their theoretical convictions.[d]

[a] So long as there is potential energy proportional to the field intensity, there will be a resultant force in the direction of maximum field gradient: qualitatively it is the existence of such forces which is set in evidence by the experiment.

[b] Cf. Mott, *An Outline of Wave Mechanics*, p. 135, 1930.

[c] See § 20.

[d] Cf. Born, *Atomic Physics*, pp. 127–30, 1935.

§ 18. *Spectral evidence.* It has been stated that the suggestion of nuclear spin moments arose first in explanation of hyperfine structure in atomic spectra. Here it will be more convenient to begin by considering the evidence from molecular spectra in certain cases. Moreover, if the suggestion of nuclear spin was not originally prompted by this evidence, some of the observations which belong to it are of an earlier date. Mecke[a] drew attention to several instances of a particular type of intensity anomaly in 1924. The change of intensity from line to line in the rotational structure of certain electronic bands of some elementary diatomic molecules was found not to be gradual or monotonic, as is usual[b] when the diatomic molecule is that of a chemical compound; on the contrary these lines were alternately "strong" and "weak" throughout the band. Occasionally, when an alternation of this type was not immediately obvious, the abnormally large separation of adjacent lines led to the suggestion[c] that the alternate "weaker" components might, in some spectra, be completely suppressed. An explanation of these effects[d] became possible with the introduction of the new mechanics. Already, on the basis of the older theory, laws of quantisation and selection rules had been evolved, which were capable of providing formally satisfactory analyses of the band spectra of a large number of diatomic molecules, more particularly of those molecules with unlike nuclei. The process of establishing these rules on the basis of the wave mechanics, with slight modification in the matter of detail, provided the opportunity for the discussion of the more important modifications which appeared necessary if the spectra of homonuclear molecules were to be included in the scheme. Heisenberg[e] and Hund[f] were the first[g] to investigate this problem;

[a] Mecke, *Phys. Z.* 25, 597, 1924; *Z. Physik,* 31, 709, 1925.

[b] Jevons, *Report on Band Spectra of Diatomic Molecules,* pp. 133–39, 1932.

[c] Mecke, *Phys. Z.* 26, 217, 1925.

[d] Jevons, *ibid.* pp. 139–44; Kronig, *Band Spectra and Molecular Structure,* § 18, 1930.

[e] Heisenberg, *Z. Physik,* 41, 239, 1927. [f] Hund, *Z. Physik,* 42, 93, 1927.

[g] Some time previously, Slater (*Nature,* 117, 555, 1926) had attempted to explain the intensity anomalies in question by a consideration of the degeneracy introduced by the identity of nuclei.

their conclusions were based upon a consideration of the symmetry properties of the wave functions describing nuclear and electronic configurations in the molecular model. Briefly, these conclusions imply that the rotational states of the molecule—and these states may be regarded as the fine structure of any electronic state (including the ground level)—must be separated into odd and even states, and the nuclear spin configurations into symmetrical and antisymmetrical configurations, and combinations made (representing actual states of the system) only[a] between symmetrical and odd, and antisymmetrical and even states, or vice versa. When the spin quantum number of each nucleus is I, symmetrical nuclear configurations are those for which the resultant spin quantum number, I_M, for the two nuclei together, takes on the values

$$2I,\ 2I-2,\ \ldots,\ 0 \text{ or } 1,$$ [b]

and antisymmetrical configurations those for which the corresponding values are

$$2I-1,\ 2I-3,\ \ldots,\ 1 \text{ or } 0.$$

Other things being equal, it may easily be shown that the total statistical weight representing all the symmetrical nuclear configurations is greater by the factor $(I+1)/I$ than the statistical weight of all the antisymmetrical configurations.[c] It is general to refer to molecules having nuclear configurations belonging to the former (symmetrical) class as ortho-molecules and to those with nuclear configurations belonging to the latter class as para-molecules. The result of the investigations of Heisenberg and Hund is, therefore, to make the distinction between the ortho and para modifications of a homonuclear diatomic molecule.

Important evidence for the significance of this distinction was almost immediately forthcoming. Dennison[d] was able to explain the apparently anomalous results regarding the low

[a] See p. 92.

[b] The ambiguity here is due to the fact that the quantum number I may be integral or half-integral.

[c] This follows at once from the result that the configuration with resultant spin quantum number I_M possesses $(2I_M+1)$-fold degeneracy: cf. Farkas (A.), *Light and Heavy Hydrogen*, chap. VI, 1935.

[d] Dennison, *Proc. Roy. Soc.* 115, 483, 1927.

temperature specific heat of gaseous hydrogen (slowly cooled from normal to low temperatures in the course of the experiment), on the assumption that no transitions from odd to even rotational states had occurred during the process. Boltzmann's distribution law appeared valid for ortho- and para-molecules separately, but not for all the molecules considered together. More direct evidence was obtained in the following year. Wood and Loomis[a] produced iodine molecules excited in a single (even) rotational level only, using monochromatic light from a mercury discharge, and observed the effect of collisions of the second kind due to the presence of helium mixed with the iodine. They found that the rotational levels excited by collision in this way were all even levels also. In 1929 the final step in confirmation was reached when Bonhoeffer and Harteck[b] succeeded in separating para-hydrogen in a relatively pure state. It may thus be regarded as experimentally established beyond doubt that transitions between ortho- and para-molecular states, as above defined, do not occur save in very exceptional circumstances. We have now to consider this fact in relation to electronic band spectra and nuclear spin.[c]

In this connection it is necessary first of all to recall the selection rule which describes the permitted change in rotational quantum number, j, in all cases. This may be written $\Delta j = 0, \pm 1$. Obviously the rotational structure of an electronic band arises precisely because of the latter possibilities, $\Delta j = \pm 1$. Furthermore, these changes are clearly changes from odd to even j, or vice versa. They are forbidden by the ortho-para conversion prohibition, unless it so happens that the electronic levels of the initial and final states of the molecule have different symmetry properties. This alone makes it possible (and then it becomes necessary)[d] that the

[a] Wood (R. W.) and Loomis (F. W.), *Phil. Mag.* 6, 231, 1928.

[b] Bonhoeffer and Harteck, *Naturwiss.* 17, 182, 1929; *Z. physikal. Chem.* B, 4, 113, 1929; see also Eucken and Hiller, *ibid.* B, 4, 142, 1929.

[c] The reader who wishes to follow other aspects of this subject is referred to Farkas (A.), *Light and Heavy Hydrogen*, part I, 1935.

[d] Whether the ortho- (para-) modification consists of molecules having even rotational states or odd depends jointly upon the symmetry properties characteristic of the electron configuration and the type of quantum statistics which is valid for the nuclei.

ortho- (para-) molecules shall have even (odd) rotational states in one stage of electronic excitation and odd (even) rotational states in the other. Thus rotational structure is not entirely absent from the band spectra of homonuclear molecules; rather is it confined to certain of the electronic bands only. When it occurs it is clear that transitions in ortho-molecules and transitions in para-molecules are each represented. Moreover, from the fact that the lines in any small region of the band follow in the order of the higher of the two j values involved in each transition,[a] it is also clear that these lines belong alternately to ortho- and para-molecules, respectively. The alternation in intensity which is observed arises in the difference in statistical weight of the two types of molecule. The ratio of these weights, $(I+1)/I$, may thus be evaluated from a comparison of the intensity of any line in respect of its neighbours on either side. In this way I, the quantum number specifying nuclear angular momentum, may be deduced. When alternate lines are missing entirely, obviously $I=0$. The intensity anomalies are completely "explained"—and, as is usual in such circumstances, the very existence of anomalies provides the means of determining an important characteristic of the physical system to which they refer. At this stage, then, it is merely necessary to discuss the advantages and disadvantages of this means of determining nuclear spin, before proceeding to the method of hyperfine structure analysis which has been more generally employed.

It has been stated that the ratio $I:I+1$ is the primary observational datum; for this reason alone the accuracy of the determination of I decreases rapidly as I increases. When $I=0$ a qualitative examination of the spectrum may give completely definite results; on the other hand intensity measurements reliable to one per cent. are necessary to distinguish between $I=4$ and $I=4\frac{1}{2}$. It is perhaps fortunate that the method is so decisive when the nuclear spin is zero, for it is precisely in such circumstances that other methods lead to ambiguous results. The great disadvantage of the method concerns the number of contingencies which may render impossible its application in a given case. Thus it is

[a] Cf. Born, *Atomic Physics*, pp. 237, 241, 1935.

necessary, not only that the species in question should be capable of existence in the (diatomic) molecular state, but also that methods of exciting the band spectrum should be available and that the spectrum so obtained should include electronic bands having rotational structure suitably isolated for investigation. With many atomic species one or other of these conditions is impossible of realisation under present conditions.

The existence of hyperfine structure in atomic spectra—according to Pauli (p. 87)—is most simply interpreted as evidence for the quantisation of the energy of magnetic interaction between the nucleus and the outer atom. According to the older viewpoint, discrete values of this energy arise in particular relative orientations of nuclear spin and resultant angular momentum of atomic electrons; these values are measurably different on account of the finite magnetic moment of the nucleus. This general basis of interpretation appears to be sufficient to account for the facts. At the outset, however, certain practical difficulties should be clearly appreciated. In the first place the fundamental effect has been described as the quantisation of the magnetic interaction energy of the atomic system in a stationary state, whereas the observational material concerns the emission of energy in the transition between one stationary state and another, and in the second place it frequently happens, with mixed elements, that complications in hyperfine structure appear, due to an isotope effect. This effect has already been mentioned (p. 86) and a further short reference to it will have to be made (p. 97).

However, once these difficulties are understood, they may be set on one side, for the present, and the matter discussed more simply in terms of the final interpretation, rather than from the point of view of the more complicated experimental facts which have given rise to it. The description of the quantisation of energy is as follows. Let us denote by I and J the quantum numbers specifying the angular momenta of the nucleus and of the extranuclear system, respectively, and by F the quantum number specifying the resultant angular momentum of the whole atom. Then, depending upon the (positive or negative) sign of the difference $(I-J)$, the possible values of F are $2J+1$ or $2I+1$ in number. They are

$$I+J, I+J-1, \ldots, |I-J|+1 \text{ and } |I-J| \qquad \ldots\ldots(19).$$

For the magnetic potential energy of the nucleus-electron interaction, similarly, $2J+1$ or $2I+1$ values are possible, namely

$$-A\mu\,[F(F+1)-I(I+1)-J(J+1)]/2IJ \qquad \ldots\ldots(20).$$

In this expression μ is the magnetic moment of the nucleus and A may be interpreted as the magnetic field in the neighbourhood of the nucleus due to the motion of the atomic electrons. So far only the simplest ideas are involved,[a] but the calculation of A in terms of the fundamental constants, e, m, h and c, and the quantum numbers specifying the electronic motions within the atom presents a different—and intricate—problem. At present the generally accepted methods of solution are those worked out by Goudsmit[b] and by Fermi and Segrè.[c] Accepting the results obtained by these investigators, (20) may be regarded as giving precisely the magnetic hyperfine structure separations (in units of energy) for a particular atomic energy state of given "total" quantum number J. Consider now two energy states between which, in the absence of nuclear spin, a radiative transition would be possible. Obviously A is different for any two such configurations of the atom; the corresponding difference in J is limited by general selection rules to the values 0 and ± 1. If, now, the further fact of nuclear spin be taken into consideration, first of all the resulting hyperfine structure multiplicity of the two states of the atom will not in general be the same[d] and, secondly, all possible transitions between the sub-

[a] The function of F, I and J in (20) is the "quantum cosine" of the angle between the directions of the axes of nuclear spin and of resultant electronic angular momentum; moreover, in respect of A, a positive value indicates that the field around the nucleus has the same direction as the latter axis (corresponding to J).

[b] Goudsmit, *Phys Rev.* 43, 636, 1933.

[c] Fermi and Segre, *Z. Physik*, 82, 729, 1933.

[d] This occurs only when $J \geqslant I$ for each state, or, if $J < I$, then when $\Delta J = 0$.

levels of one state and those of the other are no longer permitted. To this new situation the selection rule $\Delta F = 0, \pm 1$ may be shown from general principles to apply.

Here we may return, for the moment, to the data provided by experiment, in order to review exactly the conditions which must be fulfilled before the analysis of the hyperfine structure of any spectral line may be regarded as complete. From what has already been said, it is clear that the following conditions, at least, must obtain. Empirically, the frequencies of all the hyperfine structure components (or of those attributed to one particular atomic species in a mixed element) must be capable of accommodation in a system of upper and lower energy levels and to these levels must be assigned F and J values consistent with a unique value of I (cf. (19))—and with the two selection rules already given. Sometimes all this may be done without a great deal of difficulty, but in many cases a satisfactory analysis of this type is possible only from more detailed considerations concerning intensities and the relative separations of the levels.[a] As regards the latter possibility, it is evident from (20) that the energy separation between adjacent sub-levels (F and $\overline{F-1}$) is

$$-A\mu F/IJ \qquad \text{......(21)}$$

—a quantity which is proportional to F. This is Landé's interval rule: it must hold[b] for both upper and lower sets of hyperfine structure levels proposed. The intensity relations, on the other hand, are more complicated,[c] but are none the less useful in certain cases. For example, in the simplest case of all, that of the doublet hyperfine structure representing the possible transitions $J = 1$ to $J = 0$ in an atom for the nucleus of which $I = \frac{1}{2}$, the ratio of the intensities of the components is given by theory as 2 : 1.

Let it be supposed, therefore, that the term analysis of the hyperfine structure components of a given line has been successfully achieved and that quantum numbers have been

[a] Cf. Curtis, *Science Progress*, 28, 420, 1934.

[b] Instances of the failure of this rule receive separate consideration in § 21.

[c] Hargreaves, *Proc. Roy. Soc.* 124, 568, 1929; Hill, *Proc. Nat. Acad. Sci.* 15, 779, 1929.

assigned to the sub-levels involved. In this process the appropriate values of J will have been decided upon, and since the maximum value of F assigned to any sub-level of either set fixes $(I+J)$ for the set of sub-levels in question (the upper or the lower set), it is clear that I may be deduced. Then, so long as the necessary A values (expression (20)) may be calculated, the value of μ is given by simple computation. Spin quantum number and nuclear magnetic moment are thus obtained.[a] There is only one consideration which has been omitted. It has been stated (footnote, p. 95) that a positive value of A in (20) indicates a field in the same direction as the axis of resultant angular momentum of the atomic electrons. Since the charge on the electron is negative, in general negative A values must be used in the expression for the energy. Then the magnetic interaction energy increases with F amongst the sub-levels of a given set, so long as μ is positive. In actual fact the analysis of hyperfine structure components frequently issues in level systems for which this is the case ("regular" structure), but also, in other cases, level systems are deduced in which the total energy decreases as F increases ("inverted" structure). Such a state of affairs will arise, according to (20), if both A and μ are negative, in particular if the axes of the mechanical and magnetic moments of the nucleus are oppositely directed. As will be seen from Table 7, this is the generally accepted explanation of this effect. Although many attempts have been made to explain these negative μ values on the basis of a nuclear model, clearly no theory can adequately be tested until more exact numerical values are available. Nevertheless, attention may be drawn to the conclusion—generally accepted—that the neutron itself is to be described in terms of a negative μ; obviously this fact must be basic for any theory explaining negative magnetic moments in complex nuclei.[b]

So far in this account isotopic hyperfine structure has not

[a] If, in any case, μ is too small, it is clear that all hyperfine structure separations may also be too small for resolution, and thus the impression be falsely entertained that $I=0$ (cf. p. 93).

[b] See, for example, Schüler and Schmidt, *Z. Physik*, 98, 430, 1936.

been explicitly considered: it will be useful, therefore, at this stage to refer briefly to the facts[a] and theories involved, in particular to explain how it may happen that corresponding energy levels in isotopic atoms may differ slightly in energy content, although any direct effect of a nuclear mass difference alone is known to be beyond the limits of experimental detection[b] (p. 85). Roughly it may be said that the isotope displacement of corresponding levels is due to a difference in nuclear radius, rather than in nuclear mass: usually the net charge on the nucleus is distributed through a greater volume in a heavier isotope.[c] It is evident that the effect of this difference may vary enormously from one electronic state to another; broadly speaking it will be greatest for electrons "in the most penetrating orbits". Experimentally, for each level amongst a set of combining levels, for atoms of a given element, the different ("isotopic") energy values constitute a regular sequence, increasing (or decreasing) with the mass number of the nucleus. If isotopes of odd mass number be considered separately from those of even mass number this regularity is generally more pronounced still; (energy) spacings are then very nearly uniform in each series of levels—a condition which is not so evidently fulfilled for the two series taken together, since the odd numbered levels do not fall midway between the even numbered levels which bracket them.[d] It has been customary to connect this fact with the known difference in general stability (or mass defect) as between species having mass numbers odd and even, respectively.

§ 19. *The polarisation of resonance radiation.* When an atom

[a] Cf. Curtis, *loc. cit.*

[b] See, however, Dickinson, *Phys. Rev.* 46, 598, 1934. Recently, Condon also has suggested a type of explanation different from that given in the text (*Phys. Rev.* 49, 459, 1936), adopting as basis a hypothetical weak interaction between neutron and electron.

[c] Breit, *Phys. Rev.* 42, 348, 1932; *ibid.* 46, 319, 1934.

[d] In this connection the position of a level subject to *magnetic* hyperfine structure, also, is defined by the "centre of gravity" of the sub-levels by which it is replaced. It is generally the odd numbered levels which show this effect (cf. p. 107).

in the normal (ground) state absorbs energy—it may be in the form of a quantum of radiation previously emitted by an identical atom, already excited—and is thereby raised to a higher state, the radiation which it subsequently emits is referred to as resonance radiation. If the resonance radiation from a collection of atoms is excited by means of plane polarised monochromatic light, of appropriate frequency, this resonance radiation is found to be partially polarised with the same plane of polarisation as the incident light. Lack of complete polarisation may be due to several causes—operative during the finite lifetime of the atom in the excited state. If extraneous causes, such as collisions, be excluded, there is still the possibility of effects inherent in the atom itself. Thus suppose that the nucleus has non-zero mechanical and magnetic moments. According to the method of vector representation we may say that the axes of nuclear and electronic angular momentum (given by I and J, respectively) precess about the axis of their resultant at a constant rate. In particular there will be a change in the direction of the axis of total electronic moment, on this account, during the lifetime of the atom in the excited state—and the percentage polarisation of the resonance radiation will be decreased thereby. It is clear that the amount of this decrease may be reduced by the application of an external magnetic field suitably orientated. Thus an external magnetic field perpendicular to the plane of polarisation of the exciting radiation will tend to cause precession of the I and J axes of momentum around its own direction—in this case also the direction of the electric vector in the incident light—and so will tend to counterbalance the effect of precession about the randomly orientated axes of resultant atomic angular momentum (F axes). As the intensity of the applied field increases, this neutralisation becomes more pronounced until, finally, when the magnetic coupling between nucleus and atomic electrons in the excited atom has completely broken down, the polarisation effects are the same as they would be in the absence of nuclear spin. Experimentally, such effects as already described have been investigated over a considerable period;[a] the attempt to explain them in terms of nuclear moments was first made by Ellett[b] in 1930. Recently a more detailed treatment has been given by Breit.[c] Prior to this investigation, Heydenburg, Larrick and Ellett[d] had attempted to calculate I, the spin quantum number for the sodium nucleus, from measurements of the percentage polarisation of sodium-D resonance radiation in zero field, and had suggested employing similar observations in non-zero fields for the determination of μ. It is now obvious that to employ this method for the determination both of I and of μ is not generally feasible, the amount of zero field polarisation to be expected under experimental conditions being rather insensitive to the value of I. If I be known, however, the increase of polarisation with applied field intensity (H) may be employed in the determination of μ. This result follows from the fact that the amount of polarisation is strictly a function of $H/\Delta\nu$, $\Delta\nu$ being the magnetic hyperfine structure separation characteristic of the upper level involved in the radiative transition. In the notation of (20), $h\Delta\nu = A\mu$, or once $\Delta\nu$ has been determined—as a (dimensional) fitting factor between theoretical and experimental curves—μ may be calculated from this relation, if A is known. Data in respect of the resonance radiations of sodium[e] and caesium[f] have recently been treated in this way.

Closely associated with the method of polarised resonance radiation, but of rather more restricted utility, is the method of hyperfine structure in the Zeeman effect. If the applied magnetic field in these investigations be sufficient to break down completely the $I-J$ coupling, as already discussed—and not very intense fields are effective in this way—the original hyperfine structure of nuclear origin is suppressed entirely and Zeeman components are obtained precisely as if

[a] Wood (R. W.), *Phil. Mag.* 16, 184, 1908; Wood and Ellett, *Phys. Rev.* 24, 243, 1924.

[b] Ellett, *Phys. Rev.* 35, 588, 1930.

[c] Breit, *Rev. Mod. Phys.* 5, 91 (VII, § 4), 1933; see also Mitchell, *Phys. Rev.* 43, 887, 1933.

[d] Heydenburg, Larrick and Ellett, *Phys. Rev.* 40, 1041, 1932.

[e] Larrick, *Phys. Rev.* 46, 581, 1934; Ellett and Heydenburg, *ibid.* 46, 583, 1934.

[f] Heydenburg, *Phys. Rev.* 46, 802, 1934.

the nuclear spin were zero. Rather, there is only one point of difference: the Zeeman components themselves show hyperfine structure, in the form of $2I+1$ close lines[a] equally spaced. The method has been applied successfully in the case of bismuth,[b] but generally the sub-components cannot be resolved by the methods available.

§ 20. *The method of atomic beams.* This section will be concerned chiefly with the various modifications of a method suggested originally by Breit and Rabi[c] in 1931. First of all, however, reference should be made to the determinations, by Stern and his collaborators, of the magnetic moments of proton and deuteron by a molecular beam method of a different type.

Consider first the diatomic molecule H^1H^1. As previously discussed, this molecule may exist in two forms, as ortho- and para-hydrogen, respectively. In these two modifications nuclear spin axes are parallel and antiparallel and rotational quantum numbers odd and even, so that on both accounts the resultant magnetic moment of the molecule is different for the two (the molecule possesses a small magnetic moment by virtue of its rotation alone) By suitable catalysis at low temperatures pure para-hydrogen may be obtained, but at these temperatures the resultant magnetic moment of this molecule is zero on both counts. At higher temperatures, when the rotational state $j=2$ is excited, the magnetic moment is purely rotational. Under these conditions, however, pure para-hydrogen is not the equilibrium form—but in the absence of a catalyst it may be heated, when once

[a] Both the energy levels and the lines show this $(2I+1)$-fold multiplicity—on account of the selection rules which are valid in this case. Furthermore, the fine structure of the levels must still be regarded as primarily a nuclear effect, only the permitted relative "orientations" of nucleus and electron configuration are determined by their individual "setting" in the external field, rather than by their mutual interaction. With this difference the spacings of the hyperfine structure levels are given by (20) using a modified expression for the quantum cosine.

[b] Back and Goudsmit, *Z. Physik*, 47, 174, 1928. Recently Fölsche (*Naturwiss.* 24, 297, 1936) has succeeded in deducing the nuclear spin of caesium by the same method, and Green (*Phys. Rev.* 49, 866, 1936) that of Hg^{199}, likewise.

[c] Breit and Rabi, *Phys. Rev.* 38, 2082, 1931.

obtained, without transformation. Frisch and Stern,[a] therefore, determined the magnetic moment of the para-hydrogen molecule at different temperatures and so calculated the moment to be ascribed to rotation with $j=2$. They assumed that the rotational component per unit of angular momentum in ortho-hydrogen could be calculated directly from this result. On this assumption they were able to deduce the magnetic moment to be assigned to the two nuclei in ortho-hydrogen by making measurements of the effective molecular moment for normal hydrogen at a number of temperatures. There was no choice but to use normal hydrogen for this part of the investigation, since it is not possible to obtain pure ortho-hydrogen by any process of catalysis.[b] The magnetic moment of the proton was thus obtained. Quite clearly, to apply the deflection method of Stern and Gerlach in the manner described, necessitated the greatest refinements of design: the resultant magnetic moment of the ortho-hydrogen molecule at liquid air temperatures being only $\frac{1}{300}$ of that of the atom of silver, magnetic field gradients had to be increased to the utmost to obtain a final deflection so great even as 0·1 mm. But, for these refinements, the reader is referred to the original papers. The work of Estermann and Stern[c] on deuterium was based on precisely the same principles, differences in detail being due to the fact that ortho-deuterium ($\frac{5}{6}$ of these molecules have nuclear spin axes parallel, $\frac{1}{6}$ antiparallel, cf. p. 91) is the low temperature equilibrium modification in this case (possessing even j values), whilst at high temperatures it is also the more abundant (2 : 1). This fact—and the smaller value of the magnetic moment of the deuteron—added further difficulties to the determination.[d]

[a] Frisch and Stern, *Z. Physik*, 85, 4, 1933; see also Estermann and Stern, *ibid.* 85, 17, 1933.

[b] The high temperature ortho-para ratio is 3 : 1.

[c] Estermann and Stern, *Z. Physik*, 86, 132, 1933; *Nature*, 133, 911, 1934.

[d] The ratio of the magnetic moments of proton and deuteron has been calculated by Kalckar and Teller (*Nature*, 134, 180, 1934; *Proc. Roy. Soc.* 150, 520, 1935) from the observations of Farkas, Farkas and Harteck (*Proc. Roy. Soc.* 144, 481, 1934) that the rate of reconversion of para-hydrogen into normal hydrogen by paramagnetic oxygen is some sixteen times greater than that of the similarly

In describing the various methods of Breit and Rabi it is simplest to begin again with a brief reference to the original experiments of Stern and Gerlach. As already described, these experiments provide evidence for the quantisation of the additional potential energy which an atom with resultant magnetic moment possesses in an external magnetic field. At first (1922) the difficulty was to produce fields of sufficiently high gradient as to give rise to a measurable deflection of the beam. In such fields the magnetic coupling between the electron configuration and the nucleus of the atom (it is here supposed that $\mu \neq 0$) being completely destroyed (cf. p. 99), and the nuclear moment, in any case, being small compared with that due to the electrons, effectively only $2J+1$ different values of magnetic potential energy would be expected to appear. This was the interpretation placed upon the experiments at a time when nuclear magnetic moments were not seriously under consideration. Breit and Rabi pointed out that if similar deflection experiments could be made in less intense fields—such that the nucleus-electron coupling was still important—then the apparent degeneracy in the number of possible energy values would disappear and the complete set of $(2I+1)(2J+1)$ energetically different states might be set in evidence. If this could be done, obviously I might be evaluated. The first successful application of the method was described in 1933.[a] In the following year Rabi, Kellogg and Zacharias[b] deduced the mechanical and magnetic moments of proton and deuteron from observations of this type. For the exact determination of the latter moments it was necessary to produce a magnetic field uniform as regards intensity and gradient over the complete path of the atomic beam. Once more, however, the reader is referred to the original papers for the details of this achievement. The principle of the method (for the determination of μ) may be

catalysed change from ortho-deuterium to the equilibrium molecular mixture. The value so obtained ($3{\cdot}96 \pm 0{\cdot}11 : 1$) (Farkas (L.) and Farkas (A.), *Nature*, 135, 372, 1935; *Proc. Roy. Soc.* 152, 152, 1935) is in fair accord with other determinations.

[a] Rabi and Cohen, *Phys. Rev.* 43, 582, 1933.

[b] Rabi, Kellogg and Zacharias, *Phys. Rev.* 46, 157, 163, 1934.

discussed with the help of Fig. 6. Here the relative values of the component, along the direction of the field, of the effective magnetic moment of the atom are plotted as a function of the non-dimensional quantity $eH/2\pi mc . \Delta\nu$,[a] when $J=\frac{1}{2}$ and $I=1$ (e.g. for the atom H^2 in the ground state). In this connection e, m and c have their usual significance and $\Delta\nu$ is the fundamental (frequency) interval involved in the description of the magnetic hyperfine structure of the ground state of the atom. In the notation of (20), in reference to this state, $h\Delta\nu = A\mu$. Fig. 6 shows that, except in certain de-

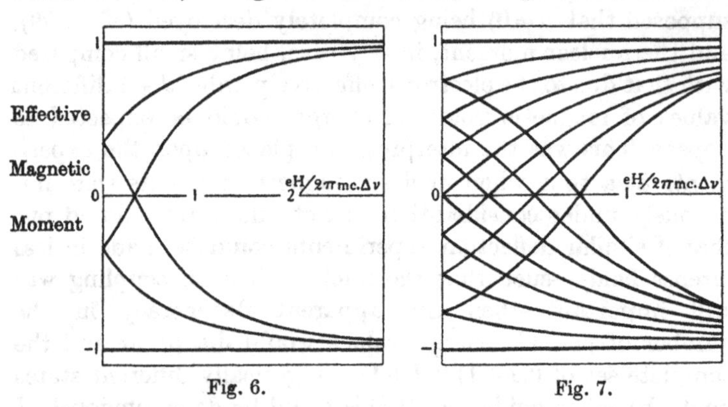

Fig. 6. Fig. 7.

generate cases, six[b] different values of magnetic moment are possible, but it is also obvious that only over a small range of field intensity, H, are these values approximately evenly spaced. The aim of experiment should be so to regulate the strength of the applied field that a deflection pattern of the greatest resolution is obtained. Relative separations having been determined for such a pattern and fitted with the predictions of Fig. 6, an experimental result of the form

$$eH/2\pi mc . \Delta\nu = x$$

enables $\Delta\nu$ to be evaluated and so μ, if A be known.

Fig. 7 has been drawn for $J=\frac{1}{2}$ and $I=\frac{7}{2}$ (e.g. for the atom

[a] This value of the parameter is appropriate when the Landé g-factor

$$[1+\{J(J+1)+S(S+1)-L(L+1)\}/2J(J+1)],$$

for the electron configuration of the atom, is 2.

[b] $(2I+1)(2J+1)=6$, in this case.

Cs^{133} in the ground state): its greater complexity allows us to discuss another variant of the atomic beam method, which was first employed by Cohen[a] in this particular case. It will be noticed that, for certain values of $eH/2\pi mc.\Delta\nu$, (doubly degenerate) states of zero effective magnetic moment are predicted. Atoms in these states will pass without deflection through the inhomogeneous field employed in the experiments. The method of "zero moments", therefore, involves the investigation of the dependence of "undeflected intensity" upon field strength. From the number of maxima so obtained, I may be found; from the values of H for which these maxima occur, μ may be calculated, as above. This method is particularly useful with mixed elements; it has already been applied on that account to the mixed elements lithium and potassium.[b]

At this stage a comparison of the hyperfine structure and atomic beam methods may be usefully made. In so far as the determination of magnetic moments is in question it is clear that the two methods are essentially similar in principle. In each case the hyperfine structure separation, $\Delta\nu$, is first obtained and a knowledge of the complete quantum specification of the atomic state is necessary before further progress may be made. The methods are different in one respect, only; in the optical method the "magnetic" structure of the levels is deduced by means of a term analysis of the corresponding structure of certain spectral lines, in the atomic beam method it is obtained directly without such analysis. It would appear, therefore, that the latter method held the advantage; on the other hand, once the term analysis of the hyperfine structure of a single line has been completed, the magnetic splitting of two energy levels has been determined. It has been a very necessary and useful check upon theory to examine for consistency the values of μ calculated from the magnetic splitting of different energy levels of the same atom. A less questionable advantage in the method of atomic beams is to be found in the fact that it is almost equally suitable—as

[a] Cohen, *Phys. Rev.* 46, 713, 1934.

[b] Millman, *Phys. Rev.* 47, 739, 1935; Fox and Rabi, *ibid.* 48, 746, 1935.

the method of zero moments—when the nuclear moment is large and when it is small: this is not true of the optical method. In that method the constant magnetic field produced by the rest of the atom is employed for the investigation of the nuclear moment; in the method of atomic beams the applied field may be varied at the discretion of the investigator. Yet hyperfine structure analysis is capable of providing one important piece of information which the other method, as hitherto employed,[a] is unable to give. By the method of atomic beams—so long as the full symmetrical pattern is employed—it is impossible to distinguish between positive and negative values of μ. This is a severe limitation.

Table 7 contains values of I and μ (in units $eh/4\pi Mc$, cf. p. 88) as at present accepted, with indications of the method or methods employed in their determination. These data are used from time to time throughout the book in the discussion of particular hypotheses; here it scarcely appears profitable to consider them further than has already been done, discursively (pp. 88, 97), during the earlier portions of this chapter.

§ 21. *Electric quadripole moments.* Hitherto the interaction between the nucleus and the outer electrons in an atom has been adequately described by attributing to the nucleus mass, electric charge, angular momentum and magnetic moment. Whether these four quantities are sufficient in all cases as a basis of explanation has, however, sometimes been questioned;[b] moreover, we have already (p. 38) discussed the attempts to explain the anomalous scattering of α particles by nuclei[c] by supposing these particles to possess a disk-like or spheroidal shape. Although these do not represent accepted theories, nevertheless they provide examples of an early recourse to quadripole moments in an attempt to explain the results of experiment. More recently, however, detailed

[a] See, however, Rabi, *Phys. Rev.* 47, 338, 1935; *ibid.* 49, 324, 1936; Kellogg, Rabi and Zacharias, *Nature*, 137, 658, 1936; *Phys. Rev.* 50, 472, 1936. The latter workers have found μ positive both for the proton and the deuteron.

[b] Cf. Racah, *Z. Physik*, 71, 431, 1931.

[c] Here, of course, nucleus-electron interaction is not in question, but a closely analogous phenomenon is involved.

Table 7

Nucleus	I	μ	Methods	Nucleus	I	μ	Methods
n^1	$\frac{1}{2}$	negative	—	Ag^{107}	$\frac{1}{2}$	−0·1	H
H^1	$\frac{1}{2}$	2·85	B, R	Ag^{109}	$\frac{1}{2}$	−0·2	H
H^2	1	0·85	B, R	Cd^{111}	$\frac{1}{2}$	−0·6	H
He^4	0	—	B	Cd^{113}	$\frac{1}{2}$	−0·6	H
Li^6	1	0·85	H, R	In^{115}	$\frac{9}{2}$	5·3	H
Li^7	$\frac{3}{2}$	3·2	B, H, R	$Sn^{117, 119}$	$\frac{1}{2}$	−0·9	H
Be^9	($\frac{1}{2}$)	—	H	Sb^{121}	$\frac{5}{2}$	2·7	H
C^{12}	0	—	B	Sb^{123}	$\frac{7}{2}$	2·1	H
N^{14}	1	<0·2	B, H	I^{127}	$\frac{5}{2}$	—	B, H
O^{16}	0	—	B	Xe^{129}	$\frac{1}{2}$	$\frac{\mu_{129}}{\mu_{131}}=1·1$	H
F^{19}	$\frac{1}{2}$	—	B, H	Xe^{131}	$\frac{3}{2}$		H
Na^{23}	$\frac{3}{2}$	2·1	B, H, R, E	Cs^{133}	$\frac{7}{2}$	2·6	H, R, E
Al^{27}	$\frac{1}{2}$	1·9	H	Ba^{137}	$\frac{3}{2}$	0·9	H
P^{31}	$\frac{1}{2}$	—	B	La^{139}	$\frac{7}{2}$	2·5	H
S^{32}	0	—	B	Pr^{141}	$\frac{5}{2}$	—	H
Cl^{35}	$\frac{5}{2}$	—	B	Eu^{151}	$\frac{5}{2}$	$\frac{\mu_{151}}{\mu_{153}}=2·2$	H
K^{39}	$\frac{3}{2}$	−0·40	B, H, R	Eu^{153}	$\frac{5}{2}$		H
K^{41}	$\frac{3}{2}$	0·22	R	Tb^{159}	$\frac{3}{2}$	—	H
Sc^{45}	$\frac{7}{2}$	3·6	H	Ho^{165}	$\frac{7}{2}$	—	H
V^{51}	$\frac{7}{2}$	—	H	Tm^{169}	$\frac{1}{2}$	—	H
Mn^{55}	$\frac{5}{2}$	—	H	Lu^{175}	$\frac{7}{2}$	1·7	H
Co^{59}	$\frac{7}{2}$	2·7	H	$Hf^{177, 179}$	$\leqslant \frac{9}{2}$	—	H
Cu^{63}	$\frac{3}{2}$	2·5	H	Ta^{181}	$\frac{7}{2}$	—	H
Cu^{65}	$\frac{3}{2}$	2·6	H	W^{183}	$\frac{1}{2}$	—	H
Zn^{67}	$\frac{3}{2}$	—	H	$Re^{185, 187}$	$\frac{5}{2}$	—	H
Ga^{69}	$\frac{3}{2}$	2·1	H	Ir^{191}	—	$\frac{\mu_{191}}{\mu_{193}} \ll 1$	H
Ga^{71}	$\frac{3}{2}$	2·7	H	Ir^{193}	$\frac{1}{2}$		H
As^{75}	$\frac{3}{2}$	0·8	H	Pt^{195}	$\frac{1}{2}$	0·6	H
$Br^{79, 81}$	$\frac{3}{2}$	—	B, H	Au^{197}	$\frac{3}{2}$	0·3	H
Se^{80}	0	—	B	Hg^{199}	$\frac{1}{2}$	0·6	H
Kr^{83}	$\frac{7}{2}$	negative	H	Hg^{201}	$\frac{3}{2}$	−0·6	H
Rb^{85}	$\frac{5}{2}$	1·4	H	Tl^{203}	$\frac{1}{2}$	1·4	H
Rb^{87}	$\frac{3}{2}$	2·9	H	Tl^{205}	$\frac{1}{2}$	1·4	H
Sr^{87}	$\frac{9}{2}$	−0·9	H	Pb^{207}	$\frac{1}{2}$	0·6	H
Nb^{93}	$\frac{9}{2}$	3·7	H	Bi^{209}	$\frac{9}{2}$	3·6	H
$Mo^{95, 97}$	$\frac{1}{2}$	—	H	Pa^{231}	$\frac{3}{2}$	—	H

Methods: B = Band spectrum, H = Hyperfine structure, R = Atomic beams (Rabi), E = Resonance radiation (Ellett).

When two nuclei are tabulated together (e.g. $Sn^{117, 119}$), it is not known whether the values of I and μ which are given refer to one or to both of these nuclei.

The table does not contain any reference to a large number of species of even mass number for which hyperfine structure has been looked for and not found. This is not necessarily to be taken as indicating that $I = 0$ for such species (cf. footnote, p. 97).

results have become available which appear to call for an explanation in some such terms. The results in question concern the validity of the Landé interval rule (p. 96) in respect of hyperfine structure of nuclear origin. The apparent failure of this rule, in itself, is no new phenomenon;[a] hitherto, however, it has always been possible to explain[b] the discrepancies in terms of certain peculiarities in the electron configurations concerned. The observations of Schüler and Schmidt[c] with europium were important because it was impossible to interpret them in this way. Therefore it was suggested that a new term in the expression for the interaction energy was required, for which the most probable physical interpretation lay in the non-spherical distribution of net positive charge in the nucleus. Casimir[d] worked out the details of this suggestion, which subsequent experimental work appears to have established as essentially correct. According to Casimir's investigation, if the distribution of charge is spherically symmetrical either in the nucleus or in the electron configuration, the effect disappears: explicitly the additional energy term in (20) may be written in the form

$$Bb \cos^2 (IJ).$$

As regards the coefficients B, b we may say that these quantities express the departure from spherical symmetry of nucleus and electron configuration, respectively, whilst, as before, $\cos^2 (IJ)$ implies the appropriate "quantum" ratio. If polar co-ordinates (r, θ) be taken, the centre of the nucleus being the pole and the nuclear spin axis the initial line, then

$$B = 2\pi \int_0^{r_0} \int_0^{\pi} \rho r^4 (3 \cos^2 \theta - 1) \sin \theta \, dr d\theta \quad \text{......(22)}.$$

Here ρ is the charge density about the point (r, θ), whilst the upper limit of integration, r_0, indicates the effectively finite radius of the nucleus. Adopting a clear-cut geometrical

[a] Schüler and Jones, *Z. Physik*, 77, 801, 1932.

[b] Casimir, *Z. Physik*, 77, 811, 1932; Goudsmit and Bacher, *Phys. Rev.* 43, 894, 1933.

[c] Schüler and Schmidt, *Z. Physik*, 94, 457, 1935.

[d] Casimir, *Physica*, 2, 719, 1935; see also Schüler, *Phys. Z.* 36, 812, 1935.

picture, it is evident that B is positive or negative as the (spheroidal) distribution of charge is elongated or flattened in the direction of the axis of spin, that is as the spheroidal nucleus is prolate or oblate. It may be remarked that, whenever the spin moments of a nucleus can be attributed to a single proton moving in the spherically symmetrical field due to the remaining constituents, it is to be expected that a negative value of B will apply. Table 8, which includes all the results[a] at present accepted, shows that positive as well as

Table 8

Nucleus	Cu^{63}	Cu^{65}	As^{75}	Eu^{151}	Eu^{153}	Lu^{175}	Hg^{201}	Bi^{209}
$q \times 10^{24}$; cm.^{2}	$-0{\cdot}1$	$-0{\cdot}1$	$0{\cdot}2$	$1{\cdot}5$	$3{\cdot}2$	$6{\cdot}1$	$0{\cdot}5$	$-0{\cdot}4$

negative values[b] are in fact obtained. Clearly, when further data are available, this information will provide an important check on any theory of nuclear constitution.

[a] Schüler and Schmidt, *Z. Physik*, 98, 430; 100, 113, 1936.
[b] If Z is the charge number of the nucleus concerned, the quantities B and q are related by the equation $B = q.Ze$.

PART THREE

CHAPTER VII

EMISSION OF α PARTICLES

§ 22. *General features of spontaneous transformations.* The discussions of § 2 have already exposed many features of the spontaneous transformations of the heavy elements, the full consideration of which is the chief concern of the science of radioactivity. Part III of this book—to which the present section must serve as introduction—is intended to summarise very briefly those portions of this science which are necessary for the understanding of the artificially produced transformations to be discussed later (Part IV): they are equally those parts of the subject which are involved in any critical survey of the problem of nuclear constitution in general. In one respect, only, Part III will have a wider content than this: the radiations from artificially produced radioelements (p. 75) will also be referred to as occasion arises (§ 26).

To begin with it will be well to recall the stage which was reached by the previous discussion. We are to accept the existence of three types of radiation and to assume a knowledge of the nature of each of them; we are in agreement[a] regarding the chance nature of spontaneous disintegration in general and have already met with the idea of the disintegration series (of parent substance and subsequent products), if only by implication (p. 38). We are in a position, therefore, to proceed to a discussion of each type of transformation in turn, but, before doing that, it will be convenient to consider, first of all, this matter of the disintegration series a little more closely. A disintegration series—according to the original usage—may be regarded as an ordered sequence of atomic species having the property that every member species except the last is radioactive, whilst the result of the disintegration of an atom of any such species is the formation

[a] The contrary view has, most recently, been put by Wolfke (*Phys. Z.* 30, 899, 1929), but now it is, less than ever, an acceptable hypothesis.

of an atom of the species next in order. From another point of view, the disintegration series may be said to exhibit the fact that the successive disintegrations of a heavy radioactive atom follow an invariable[a] sequence. For a further description of this sequence of changes some formal expression of the chance nature of the disintegration is necessary. We say that, during the relatively long periods of inactivity which separate one act of disintegration from another, the atom is characterised in turn by disintegration constants $\lambda_1, \lambda_2, \ldots, \lambda_n$ (from the macroscopic side, the decay constants of the activity of successive members of the disintegration series in question) and define these constants so as to represent, at each stage, the chance per unit time of the occurrence of the disintegration next in order.[b] This is roughly the point of view which was regarded as adequate in 1928.

Modifications became necessary for two reasons, in particular. More detailed experimental results concerning particle disintegration could no longer be reconciled with the scheme in its original simple form (many more cases of a restricted type of series branching were discovered—cf. p. 113), whilst the non-inclusion of γ radiation (radiative transformations on an equal footing with particle disintegration) became the cause for more general dissatisfaction. This arose for the following reasons: It was not contested that the quantum radiation was, in a sense, secondary to particle disintegration —either β particle or α particle disintegration—but the experiments of Ellis and Wooster[c] and of Black,[d] in the former case, and of Hahn and Meitner,[e] in the latter, showed that, in general, the time interval between the emission of the particle and the emission of the quantum was longer than the

[a] Except in so far as "series branching" occurs In 1928 four cases of this phenomenon were recognised.

[b] Cases of series branching are included by assigning partial probabilities λ_r', λ_r'' (for two alternative modes of disintegration, as the simplest case), when the effective decay constant for the corresponding product is given by $(\lambda_r' + \lambda_r'')$—whether one type of activity or the other is investigated: see Marsden and Darwin, *Proc. Roy. Soc.* 87, 17, 1912; also § 23 p. 113.

[c] Ellis and Wooster, *Proc. Camb. Phil. Soc.* 22, 844, 1925.

[d] Black, *Proc. Camb. Phil. Soc.* 22, 838, 1925.

[e] Hahn and Meitner, *Z. Physik*, 34, 795, 1925; Meitner, *ibid.* 34, 807, 1925.

time of relaxation[a] of the electrons occupying the inner electron levels in the atom. Chalfin[b] clearly expressed the opinion that, on the basis of these experiments alone, transformation constants should be regarded as applying to the radiative transformations exactly as to the particle disintegrations which precede and follow them. A further discussion was given by the author.[c] Whatever the point of view finally adopted, one result appeared certain: if γ-ray changes were to be included, as was intended, the concept of the disintegration series must lose much of its simplicity. Thus, if all spontaneous transformations (radiative and otherwise) were to be considered together (due notice being taken of differences of energy amongst the quanta emitted) it was clear that the complete sequence of changes must vary considerably from one atom to another amongst a set of radioactive atoms which are initially identical.

In the following sections—and in chap. IX, in particular—it will be our aim to show how a simple extension of the notion of the disintegration series (the inclusion of excited nuclei, along with nuclei in the ground state, as members of such a series) allows all these additional results to be comprehended.

§ 23. *Experimental data concerning the emission of α particles.* Experiment has shown (p. 15) that the α particles emitted from a radioactive preparation consisting of a mixture of radioelements possess, initially, energies constituting a line "spectrum". The earliest assumption concerning this spectrum was that a single line, only, was to be attributed to each radioelement present in the source. On that assumption the important data regarding α disintegration in any case were taken to be the energy of the particles and the disintegration constant for the change. It has been pointed out already (p. 39) that these quantities appeared to stand in a close relation one to the other.

[a] In this context, the time taken for the energy levels to respond to the change in atomic number of the nucleus on disintegration.

[b] Chalfin, *Z. Physik*, 53, 130, 1929.

[c] Feather, *Phys. Rev.* 34, 1558, 1929.

The application of the semi-circular focusing method to the determination of the velocities of these groups of apparently monoenergetic α particles (p. 18) provided the first really conclusive evidence for the failure of this simple point of view. Thus Rosenblum[a] showed that what had previously been regarded as a single group of α particles from thorium C in fact consists of a number of quite definite groups with energies distributed over a small range. Subsequent experiments revealed a similar "fine structure" in the α radiation from several radioelements; at the same time they established, with greater certainty than before, that in some cases

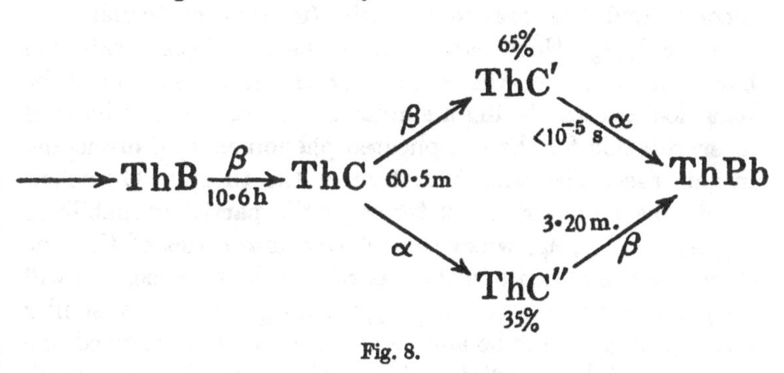

Fig. 8.

this radiation consists of one group of particles only. Thorium C′, for example, is such a body; the α radiation from thorium C′ appears to be entirely devoid of fine structure. It is instructive to pursue the comparison of the behaviour of these two thorium products somewhat farther. Let us revert again to the point of view current before Rosenblum's experiments in 1929. Then, from this point of view, a complete description of the series branching at the C product in the thorium series is provided by Fig. 8. Consider now a preparation which consists initially of pure thorium C. Almost immediately such a preparation will contain the equilibrium amount of thorium C′, the lifetime of this body being so extremely short. From the beginning, therefore, the two α particle activities of the source will steadily decrease. In 1913

[a] Rosenblum, *Comptes rendus*, 188, 1401, 1929.

Marsden and Wilson[a] showed that the ratio of the numbers of α particles constituting these two activities remained to all appearances constant as such a source decayed. At a time when the complexity of the α radiation from thorium C was not suspected, this result was regarded as establishing the interpretation already noted (footnote, p. 111); the transformation of the thorium C nucleus is to be described in terms of partial disintegration constants (λ_α, λ_β) in such a way that the probabilities (per unit time) of the (mutually exclusive) α and β modes of disintegration are λ_α and λ_β, respectively. The chance of one or other transformation is then $(\lambda_\alpha + \lambda_\beta)$ per second, and the branching ratio (in this particular case 35/65) is $\lambda_\alpha/\lambda_\beta$. Now that the fine structure of the α radiation has been detected, the same experimental result must be regarded as establishing a similar but more detailed basis of interpretation for the complicated phenomenon of branching in this case. The transformation of the thorium C nucleus must now be described in terms of the partial probabilities $\lambda_{\alpha 1}$, $\lambda_{\alpha 2}$, ... $\lambda_{\alpha 6}$, λ_β, when the relative intensities of the fine structure components in the α particle velocity spectrum will be given by the ratios $\lambda_{\alpha 1} : \lambda_{\alpha 2} : \ldots : \lambda_{\alpha 6}$. Precisely similar formal schemes may be adopted to describe the observed fine structure of the α radiation from other radioelements which do not show series branching of the more obvious (α/β) type: in each case the relative intensities of the fine structure components remain constant as the general α activity decays.

At this stage, when partial disintegration constants have been defined, it may be noted how the simplicity of the original notions concerning α disintegration has been further destroyed by the discovery of Rosenblum. In certain cases the fine structure component of the highest energy is not the most intense; quite clearly the Geiger-Nuttall rule in its simple form is no longer valid in these cases—a decrease in the energy of disintegration is accompanied by an increase in the transformation probability. We shall return to this point again later (p. 148).

Table 9 contains disintegration constants and disintegra-

[a] Marsden and Wilson, *Phil. Mag.* 26, 354, 1913.

Table 9. α radiation from nuclei in the ground state

Radioelement	(Partial) disintegration constant: sec.$^{-1}$	Disintegration energy: electron volts × 10^{-6}	Percentage intensity
U_I	$4{\cdot}8 \times 10^{-18}$	4·31	
U_{II}	$1{\cdot}3 \times 10^{-13}$	4·89	
Io	$2{\cdot}6 \times 10^{-13}$	4·84	
Ra	$1{\cdot}35 \times 10^{-11}$	4·879	98
	$0{\cdot}03 \times 10^{-11}$	4·695	2
Rn	$2{\cdot}10 \times 10^{-6}$	5·589	
RaA	$3{\cdot}8 \times 10^{-3}$	6·112	
RaC	$5{\cdot}5 \times 10^{-8}$	5·612	45
	$6{\cdot}6 \times 10^{-8}$	5·550	55
RaC′	$3{\cdot}5 \times 10^{3}$	7·829	
RaF	$5{\cdot}7 \times 10^{-8}$	5·403	
Pa	$6{\cdot}9 \times 10^{-13}$	5·24	
RdAc	$10{\cdot}2 \times 10^{-8}$	6·159	24
	$0{\cdot}6 \times 10^{-8}$	6·127	1·5
	$8{\cdot}1 \times 10^{-8}$	6·097	19
	$0{\cdot}4 \times 10^{-8}$	6·075	1
	$1{\cdot}3 \times 10^{-8}$	6·030	3
	$1{\cdot}9 \times 10^{-8}$	5·975	4·5
	$1{\cdot}3 \times 10^{-8}$	5·921	3
	$9{\cdot}3 \times 10^{-8}$	5·869	22
	$0{\cdot}4 \times 10^{-8}$	5·847	1
	$7{\cdot}7 \times 10^{-8}$	5·822	18
	$1{\cdot}3 \times 10^{-8}$	5·776	3
AcX	$3{\cdot}0 \times 10^{-7}$	5·823	42
	$2{\cdot}4 \times 10^{-7}$	5·709	34
	$1{\cdot}2 \times 10^{-7}$	5·634	17
	$0{\cdot}5 \times 10^{-7}$	5·54	7
An	$1{\cdot}22 \times 10^{-1}$	6·953	69
	$0{\cdot}27 \times 10^{-1}$	6·683	15
	$0{\cdot}21 \times 10^{-1}$	6·556	12
	$0{\cdot}07 \times 10^{-1}$	6·34	4
AcA	$3{\cdot}5 \times 10^{2}$	7·508	
AcC	$4{\cdot}5 \times 10^{-3}$	6·739	84
	$0{\cdot}8 \times 10^{-3}$	6·383	16
AcC′	?	7·581	
Th	$1{\cdot}7 \times 10^{-18}$	4·16	
RdTh	$9{\cdot}7 \times 10^{-9}$	5·517	84
	$1{\cdot}9 \times 10^{-9}$	5·431	16
ThX	$2{\cdot}2 \times 10^{-6}$	5·786	
Tn	$1{\cdot}27 \times 10^{-2}$	6·400	
ThA	5·0	6·904	
ThC	$1{\cdot}83 \times 10^{-5}$	6·201	27·2
	$4{\cdot}65 \times 10^{-5}$	6·161	69·7
	$0{\cdot}12 \times 10^{-5}$	5·873	1·8
	$0{\cdot}01 \times 10^{-5}$	5·728	0·16
	$0{\cdot}07 \times 10^{-5}$	5·709	1·1
	10^{-6}	5·572	0·01
ThC′	?	8·948	

tion energies[a] for all α bodies in the three main series—and, in addition, for further convenience, the percentage intensities of the sub-groups in those cases in which fine structure has been established.[b]

So far nothing has been said regarding "long range" α particles. Already in 1916 Rutherford and Wood[c] had discovered a very weak component of high energy (possessing about one ten-thousandth of the intensity of the main group) in the α radiation from thorium active deposit, and three years later Rutherford[d] found a similar, though still feebler, component in the radiation from the active deposit of radium. At that time, owing to the relatively small intensities of these high energy radiations, not much attention was devoted to them. Recently, however, they have been very thoroughly studied,[e] with most interesting results. The interpretation of these results[f] to be offered in chap. IX will be found to justify the segregation of the data referring to them in a separate table. Table 10 contains the disintegration energies and the intensities, with respect to the normal mode of disintegration, for each of the abnormal types of disintegration which give rise to α particles of long range. In view of the interpretation which is to be given (p. 146), it may be well to remark that the relative intensities in this table are merely those directly determined with a source of active deposit; they are not proportional (for each product) to the corresponding partial disintegration constants for the abnormal modes of disintegration—as is the case with the fine structure groups (Table 9).

[a] The disintegration energy here given is the sum of the kinetic energies of the α particle and the recoiling nucleus: see Rutherford, Ward and Lewis, *Proc. Roy. Soc.* 131, 684, 1931—also Enskog, *Z. Physik*, 45, 852, 1927.

[b] This table has been compiled chiefly from the data given by Lewis and Bowden (*Proc. Roy. Soc.* 145, 235, 1934). For the specific charge of the α particle at rest these authors adopted the value 4822·3 e.m.u. per gm. and, for the velocity of light, $2{\cdot}9980 \times 10^{10}$ cm. per sec.; see equation (29) p. 139.

[c] Rutherford and Wood, *Phil. Mag.* 31, 379, 1916.

[d] Rutherford, *Phil. Mag.* 37, 571, 1919.

[e] For references to the earlier work see Nimmo and Feather, *Proc. Roy. Soc.* 122, 668, 1929; for the most recent results, Rutherford, Lewis and Bowden, *ibid.* 142, 347, 1933.

[f] This interpretation is anticipated, to some extent, in the descriptive headings of Tables 9 and 10.

Table 10. α radiation from nuclei in various states of excitation

Radioelement	Disintegration energy: electron volts $\times 10^{-6}$	Relative intensity $\times 10^{6}$
RaC′	8·437	0·43
	9·112	0·45
	9·242	22
	9·493	0·38
	9·673	1·35
	9·844	0·35
	9·968	1·06
	10·097	0·36
	10·269	1·67
	10·342	0·38
	10·526	1·12
	10·709	0·23
ThC′	9·674	34
	10·745	190

§ 24. *Theoretical considerations.* The only aspect of theory here to be considered is just such an elaboration of the ideas of § 10 as is sufficient to form a basis for further discussion (chap. IX). We need to show (employing the notion of the potential barrier, already developed) how disintegration probabilities may be calculated in terms of the energy of disintegration and other relevant "constants". It is clear, first of all, that the value of the nuclear charge must be one of these constants. This charge[a] determines the form of the potential barrier at distances from the centre which are at least somewhat greater than the distance of the summit. Then, one other constant, if not more, must be added to describe the form of the barrier around and within the summit. Generally two such constants have been employed, approximation to the continuous form, $U(r)$, of the potential energy curve of Fig. 1 being effected by cutting off the curve of potential energy corresponding to a pure inverse square law repulsion abruptly at a distance r_0 and assigning a constant value, $-U_0$, to the potential energy within this dis-

[a] Strictly, the charge on the product nucleus—cf. footnote, p. 49.

tance. It should be remarked at once that such a method of approximation might be quite unsuited to the treatment of some problems (it is not a good approximation if we wish to calculate the proper energies corresponding to the virtual energy levels for α particles in the nucleus (p. 51) or the nature of the anomalous scattering of such particles by it (p. 49)), but it turns out to be a good one for the purpose in hand. In actual fact the values obtained for disintegration probabilities do not differ greatly according to the assumptions which are made regarding the inner attractive field of the nucleus—provided only that the field is not large except in a narrow region around the nucleus[a] (barrier with steep inner wall). If E is the energy of the emitted particle, of mass M, the important factor in determining this probability may be shown to be[b]

$$e^{-\frac{4\pi (2M)^{\frac{1}{2}}}{h}\int_{r_1}^{r_2}\{U(r)-E\}^{\frac{1}{2}}dr} \quad \text{......(23)},$$

where the limits of integration, enclosing that range of values of r for which the integrand is real, also represent the boundaries of the region in which, according to the classical interpretation, the kinetic energy of the particle is negative. The approximate independence of this factor of the inner form of the barrier, within the latitude which the above conditions allow, is here immediately obvious.

As already indicated $\{E-U(r)\}$ in (23) represents the kinetic energy of the α particle at a distance r from the centre of the nucleus—and, in this connection, until now we have regarded the motion of the particle as being purely a radial one. When disintegration takes place with the formation of a nucleus differing in angular momentum from the original nucleus, however, a simplification of this kind is no longer valid; the balance of momentum must be regarded as taken up by the particle which is emitted. If the nuclear spin

[a] This result, and the very nearly regular variation of α disintegration energy with atomic number in a given radioactive series, provides the reason for the fact that disintegration constants depend as nearly exclusively upon energy of disintegration as the Geiger-Nuttall rule suggests.

[b] Cf. Gamow, *Structure of Atomic Nuclei and Nuclear Transformations*, 1936, chap. V.

quantum number changes by j units in the process of disintegration, angular momentum $\{j(j+1)\}^{\frac{1}{2}} h/2\pi$ will be carried by this particle. With such angular momentum, at a distance r from the centre of force, kinetic energy of transverse motion equal to $j(j+1)\, h^2/8\pi^2 Mr^2$ must obviously be associated. Clearly the kinetic energy of radial motion is then

$$E - U(r) - j(j+1)\frac{h^2}{8\pi^2 Mr^2} \qquad \ldots\ldots(24),$$

instead of $\{E - U(r)\}$, when $j = 0$. Making the appropriate substitution in (23), it will be seen that, for a given energy, disintegration constants are all materially reduced. It is as if, in these cases (in which $j \neq 0$), the potential barrier were raised by the superposition of a barrier of potential energy "corresponding to centrifugal force". Only one difficulty remains: the third term in (24) becomes infinite when r tends to zero, or the new potential barrier rises continuously towards the origin. Probably a difficulty of this kind is inherent in the general method of attack—we cannot, strictly, describe the disintegration in terms of x particle and residual nucleus as distinct entities capable of separate consideration throughout—but it may be mentioned that numerical results having a close bearing upon actual fact (cf. § 31) have been obtained by cutting off this additional potential barrier, along with that representing the forces of electrostatic repulsion, at the critical distance r_0, as above described. Then, putting in accepted values for the fundamental constants and remembering, in respect of the term in j, that we are concerned with elements for which the atomic number, Z, is of the order of 100, (25) may be obtained as an expression for λ in a convenient numerical form:

$$\log_{10}\lambda = -2{\cdot}40 - 2\log_{10} r_0 - 1{\cdot}727 \times 10^3 (Z-2)\, V^{-\frac{1}{2}} + 4{\cdot}097 \times 10^6 (Z-2)^{\frac{1}{2}}\, r_0^{\frac{1}{2}} \{1 - 10^{-3}.j(j+1)\} \qquad \ldots\ldots(25).$$

Whilst, previously, E has been employed to denote the energy of the particle in ergs, V is here used to refer to the same energy in electron volts.[a] From what has already been said, the absence of U_0 from (25) will occasion no surprise, but it will be seen that r_0 remains—as the one arbitrary constant by means of which to achieve agreement between theory and experiment. This agreement may be satisfactorily obtained by giving r_0 values increasing slightly with Z, for the radioelements of any disintegration series, and of absolute magnitude of the order of 8×10^{-13} cm. This adjustment would appear to be entirely reasonable.

[a] Strictly, E in (23) and V in (25) refer not to the energy of the α particle, or to the disintegration energy, exactly, but, if the former be denoted by E' and the latter by E'', we have, as accurately as need be, $E = E'A^2/(A-4)^2 = E''A/(A-4)$, and similarly for V. Here A is the mass number of the nucleus before disintegration.

CHAPTER VIII

EMISSION OF ELECTRONS, POSITIVE AND NEGATIVE

§ 25. *The emission of β particles by naturally radioactive substances.* It has already been mentioned (p. 41) that the primary β particles emitted by different disintegrating atoms of a particular radioelement may vary widely in energy. In all cases so far investigated the distribution of energy amongst the disintegration particles appears to be a continuous one—with a definite upper limit and a single[a] maximum (for an energy certainly less than half the limiting energy). Yet, as in the α particle case, it is evident, also, that a unique disintegration constant, λ, describes the probability of nuclear transformation per unit time: however it may be measured the decrease in activity of a pure β particle preparation is found to be simply exponential.[b] We may therefore accept the complete energy distribution amongst the primary β particles from such a source as constituting, along with the appropriate disintegration constant, the important experimental data with respect to that source. By a natural extension of the formalism adopted in describing the phenomena of branching (pp. 111, 113) it is possible to combine these data, for a given element, in a single statement. We may say that the radioactive nucleus is characterised by a continuous range of mutually exclusive (infinitesimal) probabilities of disintegration, in such a way that the chance of the emission of a β particle with energy between E and $E+\delta E$, in any time δt (at the beginning of which the nucleus is still unchanged), is

$$\Lambda(E)\,\delta E\,.\,\delta t.$$

[a] There is still some uncertainty concerning the relative numbers of primary electrons having very small energies: see Richardson (H. O. W.), *Proc. Roy. Soc.* 147, 442, 1934; Scott, *Phys. Rev.* 48, 391, 1935; Alichanow, Alichanian and Dzelepow, *Nature*, 137, 314, 1936.

[b] Bastings, *Phil. Mag.* 48, 1075, 1924; Sargent, *Trans. R.S. Canada*, III, 26, 205, 1932.

If the limiting energy of the disintegration electrons is E_0, $\Lambda(E)$ is finite only in the range $0 < E < E_0$, and λ is given by the relation

$$\lambda = \int_0^{E_0} \Lambda(E)\, dE \qquad \text{......(26).}$$

From the alternative, macroscopic, point of view, of the primary electrons emitted, in any interval of time from a pure source of the radioelement in question, the fraction having kinetic energies between E and $E + \delta E$ is, clearly,

$$\lambda^{-1} . \Lambda(E)\, \delta E.$$

If this fraction is denoted by $F(E)\, \delta E$, the function F may be referred to as the energy distribution function for the β particle transformation under discussion (cf. Fig. 9). At a later stage (p. 149) it may be convenient to analyse $F(E)$ into partial distribution functions, in certain cases, but it must be emphasised here that there is no direct experimental method of making any such analysis: all evidence for its significance must be drawn from quite other data.

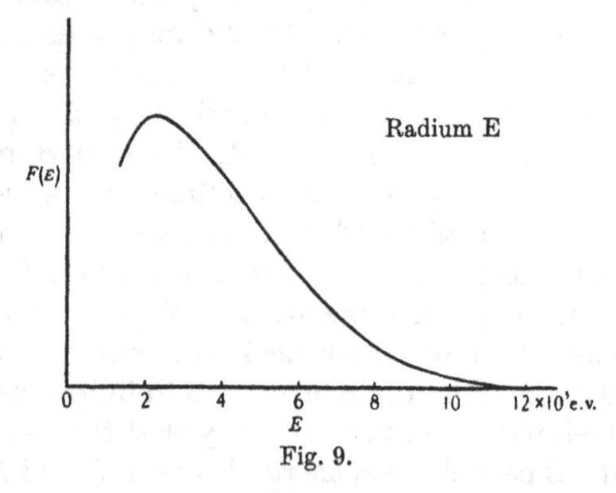

Fig. 9.

Table 11 contains the disintegration constants[a] and, when these are known, the limiting energies of the β particles from the various β active bodies of the three main series; as some indication of the form of the function F, it includes also the ratio E_m/E_0, of the most probable energy, E_m, to the maximum energy of the particles which are emitted. On this

[a] Partial disintegration constants are employed when α/β branching occurs.

evidence it would appear that, contrary to earlier speculations, the energy distribution function cannot be of the same form in all cases. On the other hand it is important not to overlook the fact that incomplete experimental evidence is the rule rather than the exception in this domain.[a]

Table 11

Radioelement	λ sec.$^{-1}$	E_0 e.v. $\times 10^{-6}$	E_m/E_0
UX_1	$3\cdot3 \times 10^{-7}$	$\sim0\cdot15$	—
UX_2	$1\cdot0 \times 10^{-2}$	$2\cdot32$	$0\cdot29$
UZ	$2\cdot9 \times 10^{-5}$	$\sim0\cdot6$	—
RaB	$4\cdot31 \times 10^{-4}$	$0\cdot65$	$0\cdot26$
RaC	$5\cdot86 \times 10^{-4}$	$3\cdot15$	$0\cdot13$
RaC″	$8\cdot7 \times 10^{-3}$	>2	—
RaD	$1\cdot0 \times 10^{-9}$	$<0\cdot04$	—
RaE	$1\cdot6 \times 10^{-6}$	$1\cdot1$	$0\cdot2$
UY	$7\cdot8 \times 10^{-6}$	$\sim0\cdot2$	—
Ac	$1\cdot6 \times 10^{-9}$	$0\cdot22$	—
AcB	$3\cdot21 \times 10^{-4}$	$0\cdot3$	—
AcC	$1\cdot7 \times 10^{-5}$	—	—
AcC″	$2\cdot43 \times 10^{-3}$	$1\cdot4$	$0\cdot24$
$MsTh_1$	$3\cdot3 \times 10^{-9}$	—	—
$MsTh_2$	$3\cdot14 \times 10^{-5}$	$2\cdot05$	—
ThB	$1\cdot82 \times 10^{-5}$	$0\cdot36$	$0\cdot36$
ThC	$1\cdot24 \times 10^{-4}$	$2\cdot25$	$0\cdot16$
ThC″	$3\cdot7 \times 10^{-3}$	$1\cdot80$	—

Yet, in spite of this fact, a number of important generalisations has been suggested on the basis of this evidence. Thus, at an early stage, Gurney[b] remarked, in comparing the energies of the β particles and the γ rays emitted by the same radioelement, that, in general, the limiting energy of the particles (E_0) was somewhat greater than the maximum quantum energy of the radiation. Present theory provides an interpretation of this connection (p. 149), as, also, of its most glaring exception (the case of thorium C″). Secondly, there is the generalisation that the energy of the upper limit, E_0,

[a] See Sargent, *Proc. Camb. Phil. Soc.* 28, 538, 1932; *Proc. Roy. Soc.* 139, 659, 1933.

[b] Gurney, *Proc. Roy. Soc.* 112, 380, 1926.

should be regarded as the disintegration energy[a] for the transformation in question. Experiments which have already been mentioned (p. 41) appeared to cast doubt on this

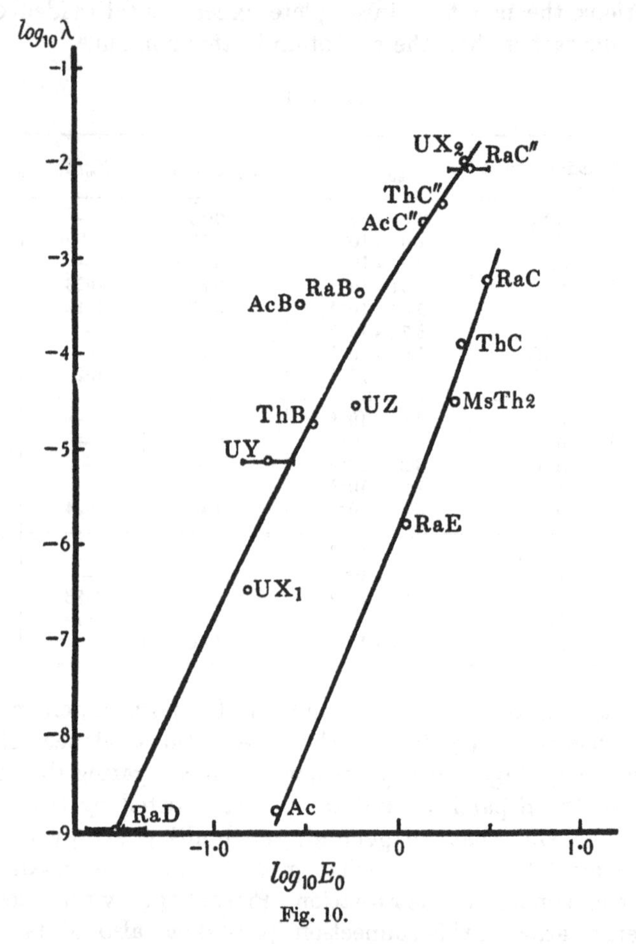

Fig. 10.

contention: the attempt to prove the emission of an amount of energy E_0, in some form or other, from each disintegrating atom resulted in failure. As against this, however, were two results which carry greater conviction. One of these was of a

[a] With β particles there is no need to make the calculation for nuclear recoil.

general nature, the other concerned a particular case. Sargent[a] put forward the general result; he showed that, if limiting energies, E_0, were employed, a numerical relation, of the Geiger-Nuttall type, connected decay constant and energy of disintegration. This is clear from Fig. 10. Originally the division of β-active substances into two groups, as suggested by the diagram, was accepted as an empirical result; the theory of β disintegration, however, affords a simple basis of explanation (p. 135). The particular result, in its final form, follows from the accurate observations of Henderson.[b] Henderson established the fact that alternative modes of disintegration of thorium C (see Fig. 8) take place with the same release of energy, only if β transformation energies E_0 are assumed. The experimental values (taken from Tables 9 and 11 and from p. 149) are as follows (electron volts $\times 10^{-6}$):

C – C′ branch. α particle, 8·95; β particle, 2·25; γ rays, 0; total 11·20.
C – C″ branch. α particle, 6·20; β particle, 1·80; γ rays, 3·20; total 11·20.

It will be observed that the agreement is exact. It became necessary, therefore, to renew the search for the "missing" energy—or, at least, to suggest hypotheses which would "explain" its non-appearance (p. 134).

§ 26. *Positive and negative electrons from artificially produced radioelements.* Experiments on artificial disintegration—until quite recently—had always been negative in two respects: they had never demonstrated the emission of high-energy electrons as disintegration products[c] and they had failed to produce any evidence for the persistence of transformation processes after the removal of the source of bombarding particles.[d] Positive results in the latter of these two particulars followed very closely upon similar results in the former. In 1933 Curie and Joliot[e] found that both positive and

[a] Sargent, *Proc. Roy. Soc.* 139, 659, 1933.
[b] Henderson (W. J.), *Proc. Roy. Soc.* 147, 572, 1934.
[c] Cf. Bothe, *Phys. Z.* 32, 661, 1931.
[d] See Shenstone, *Phil. Mag.* 43, 938, 1922.
[e] Curie and Joliot, *Comptes rendus*, 196, 1885, 1933; *J. Physique*, 4, 494, 1933.

negative electrons were emitted, along with heavy particles, from thin layers of beryllium, boron and aluminium bombarded by the α particles from polonium. From certain differences observed it appeared that the phenomenon in the case of beryllium was not of the same nature as with the other two elements. The positrons from boron and aluminium were ascribed to nuclear, those from beryllium to extranuclear processes (cf. p. 132). In 1934 this interpretation was upheld and the nature of the nuclear processes made clear through the discovery,[a] by the same workers, of the persistence of some positron emission after removal of the polonium. It was found, in fact, that the positron activity of irradiated boron (or aluminium) decreased exponentially with increasing time after the end of the exposure. Similarly, during exposure, this activity rose from zero, initially, to a limiting value. These results may clearly be explained by assuming the production, during the primary bombardment, of a radioactive species of short life, differing only from the natural radioelements in its low atomic number and the fact that positrons, rather than electrons, are emitted in its disintegration. The activities from boron and aluminium were in this way ascribed to previously unknown species ${}_{7}N^{13}$ and ${}_{15}P^{30}$, respectively. It was not long before confirmatory chemical evidence was obtained for these assumptions:[b] inactive (ordinary) nitrogen or phosphorus being added to the activated substance, as the case may be, it was found that the new activity could be collected with the added material by means of any suitable chemical separation.

This discovery of "artificial" (positron) radioactivity was rapidly confirmed and extended, and, with at least one element (magnesium), negative electron activity was also established—as the result of α particle bombardment.[c] Then artificial radioactivity (both negative and positive electron activity) was produced using other types of projectile—

[a] Curie and Joliot, *Comptes rendus*, 198, 254, 1934.

[b] Curie and Joliot, *Comptes rendus*, 198, 559, 1934; *J. Physique*, 5, 153, 1934.

[c] Curie and Joliot, *J. Physique*, 5, 153, 1934; Alichanow, Alichanian and Dzelepow, *Nature*, 133, 871, 1934.

protons,[a] deuterons[b] and neutrons.[c] The many primary reactions involved are discussed in the appropriate sections of Part IV; here data concerning the subsequent radioactivity, alone, are of interest. A selection[d] of these is given in Tables 12 and 13—the data concerning negative-electron-active substances in the former table, that referring to positron-

Table 12

Radio-element	τ	E_0 e.v. $\times 10^{-6}$	Radio-element	τ	E_0 e.v. $\times 10^{-6}$
Li^8	0·5 s.	10	$In^{112, 114, 116}$	13 s.	—
B^{12}	0·02 s.	11		54 m.	—
C^{14}	~90 d.	0·2		3·5 h.	—
N^{16}	8·5 s.	7	I^{128}	25 m.	—
O^{19}	40 s.	—	Cs^{134}	1·5 h.	—
F^{20}	12 s.	5·0	La^{140}	1·9 d.	—
Ne^{23}	40 s.	—	Pr^{142}	19 h.	—
Na^{24}	15·5 h.	1·85	Tb^{160}	3·9 h.	—
Mg^{27}	10 m.	2·05	Lu^{176}	3·8 h.	—
Al^{28}	2·3 m.	3·3	Hf^{181}	>100 d.	—
Al^{29}	11 m.	>3	W^{187}	1 d.	—
Si^{31}	2·5 h.	1·85	Ir^{194}	50 m.	—
P^{32}	15 d.	1·6	Au^{198}	2·6 d.	0·3
S^{35}	>60 d.	—	Hg^{205}	1·7 d.	—
Cl^{38}	35 m.	4·8	$Tl^{204, 206}$	4 m.	—
A^{41}	1·8 h.	2·7		1·6 h.	—
K^{42}	16 h.	3·5	Ra^{229}	1 m.	—
V^{52}	3·75 m.	—	$Ac^{229, 232}$	12 m.	—
Mn^{56}	2·5 h.	—		3·5 h.	—
$Cu^{64, 66}$	5 m.	—	Th^{233}	25 m.	—
	10 h.	—	Th^{235}	4 m.	—
Ga^{70}	20 m.	—	Pa^{233}	2·5 m.	—
As^{76}	1·1 d.	1·5	Pa^{235}	v. short	—
$Br^{78, 80, 82}$	18 m.	—	U^{235}	24 m.	—
	4·2 h.	—	U^{237}	40 s.	—
	1·5 d.	—	U^{239}	15 s.	—
Y^{90}	2·9 d.	—	$EkaRe^{237}$	16 m.	—
Rh^{104}	40 s.	3·6	$EkaRe^{239}$	2·2 m.	—
$Ag^{108, 110}$	25 s.	3·8	$EkaOs^{237}$	12 h.	—
	2·5 m.	2·8	$EkaOs^{239}$	59 m.	—

[a] Cockcroft, Gilbert and Walton, *Nature*, 133, 328, 1934.
[b] Lauritsen, Crane and Harper, *Science*, 79, 234, 1934.
[c] Fermi, *La Ricerca Scientifica*, 5(1), 283, 1934.
[d] Table 12 has reference only to those cases in which the identity of the radioactive species is known with some certainty. Many other examples of the phenomenon, in which the half-value period and the chemical nature of the radioactive substance have been determined, are still under investigation. Since, however, mass numbers are unknown, these cases are not included in the table.

active substances in the latter. As before, the disintegration electrons (and positrons) have energies which constitute a continuous spectrum[a] in each case; on the previous supposition that the upper limit of energy in this spectrum is the important datum, this limiting energy is listed (as E_0) in the tables. In column 2, τ is the half-value period of the corresponding activity.

Table 13

Radio-element	τ	E_0 e.v. $\times 10^{-6}$	Radio-element	τ	E_0 e.v. $\times 10^{-6}$
C^{11}	21 m.	1·15	Si^{27}	8 m.	1·5
N^{13}	10·5 m.	1·25	P^{30}	3·2 m.	2·9
O^{15}	2·1 m.	1·7	(S^{31})	2·1 d.	0·9
F^{17}	1·15 m.	2·1	Cl^{34}	40 m.	—
Na^{22}	?	—	Sc^{43}	4·4 h.	—
Al^{26}	7 s.	—	Sc^{44}	3 h.	—

A most interesting result is indicated in Fig. 11. Here logarithms of disintegration constants are plotted against logarithms of the limiting energies, E_0, just as in Fig. 10, the two smooth curves of that figure being repeated for sake of reference. It will be seen that the points[b] representing the behaviour of the radioactive elements of low atomic number fall, along with those referring to the heavier elements, surprisingly well[c] in the region traversed by the two curves. Clearly, the disintegration constant for β transformation does not vary rapidly with atomic number (for a given energy of disintegration) as appears to be the case for the α particle change (§ 24). Again, there is no great difference between the probabilities of electron and positron emission, for the same release of energy.

One further consideration must suffice. It will be observed that Table 12 contains data in respect of 16 radioactive species

[a] Curie and Joliot, *J. Physique*, 5, 153, 1934; Ellis and Henderson, *Proc. Roy. Soc.* 146, 206, 1934.

[b] In one or two cases the points of Fig. 11 and the data of Table 12 do not exactly correspond. In these cases later experimental material has been included in the table without the appropriate change having been made in Fig. 11.

[c] It is interesting to remark that the chief discrepancies occur for the species P^{32}, Cl^{38} and K^{42}—the heaviest known isotopes of the elements having consecutive odd atomic numbers 15, 17 and 19.

of atomic number greater than 80, that is 16 species either isotopic with one or other of the naturally occurring radio-

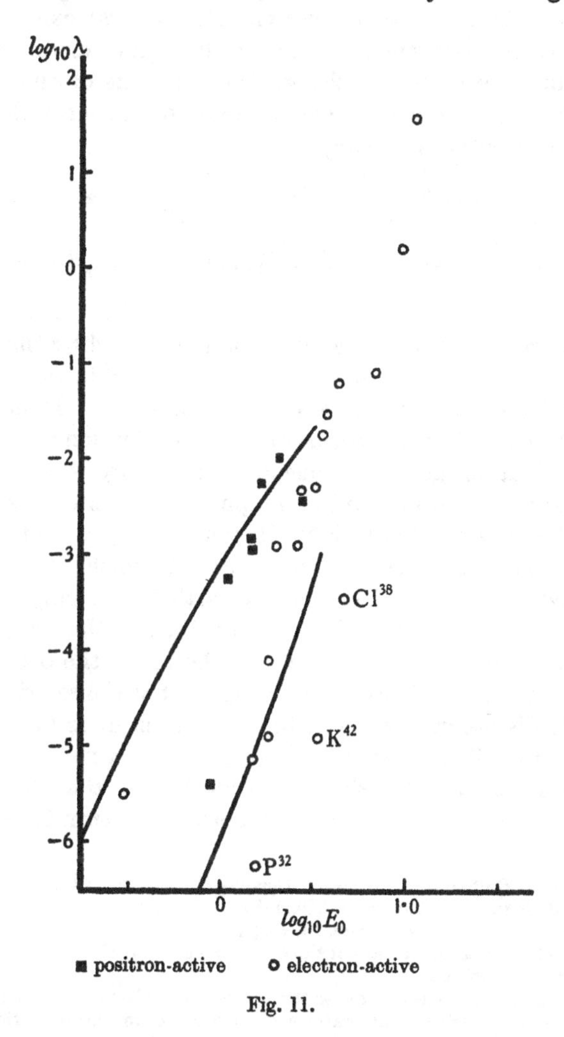

Fig. 11.

elements of the three main series, or of still higher atomic number.[a] Most of these species have been discovered as

[a] Fermi, *Nature*, 133, 898, 1934.

products resulting from the bombardment of thorium[a] and uranium[b] with neutrons. They are interesting in many respects. In the first place evidently several examples of successive transformation occur: rudimentary disintegration series may be traced in the various relations of one species with another. Of such sequences perhaps the most clear-cut examples are the following:

$$ {}_{90}\mathrm{Th}^{233} \xrightarrow[25\text{ m.}]{\beta} {}_{91}\mathrm{Pa}^{233} \xrightarrow[2\cdot5\text{ m.}]{\beta} {}_{92}\mathrm{U}^{233} \xrightarrow{?} \qquad \ldots\ldots(27)$$

$$ {}_{92}\mathrm{U}^{239} \xrightarrow[15\text{ s.}]{\beta} {}_{93}\mathrm{EkaRe}^{239} \xrightarrow[2\cdot2\text{ m.}]{\beta} {}_{94}\mathrm{EkaOs}^{239} \xrightarrow[59\text{ m.}]{\beta} {}_{95}\mathrm{EkaIr}^{239} \xrightarrow{?} \qquad \ldots\ldots(28).$$

This is precisely[c] the type of relation exhibited by the radioelements of the main series. Secondly, certain of the new species have mass numbers of the type $4n+1$. It has long been recognised that a series of this specification[d] was to be expected from general considerations, since the thorium ($4n$), uranium ($4n+2$) and actinium ($4n+3$) series exemplify the other possibilities.[e] Now it would appear that (27) is part of such a series, the members of which, it must be assumed, no longer exist upon the earth to-day only because of the relatively short lifetime of their ultimate parent, whatever species that may be. Furthermore, the occurrence of the sequence (28) would also suggest that the restriction of possible disintegration series to one of each mass type is not really admissible; (28) is part of a second series with mass numbers $4n+3$. In these considerations the distincton between natural and artificial radioelements stands revealed in its flimsiest extreme.[f]

[a] Curie, v. Halban and Preiswerk, *Comptes rendus*, 200, 1841, 2079, 1935; Hahn and Meitner, *Naturwiss*. 23, 320, 1935.

[b] Meitner and Hahn, *Naturwiss*. 24, 158, 1936.

[c] Except in so far as no case of three successive β particle changes had previously been recognised.

[d] So long as no particle other than the α particle and the electron is emitted in a sequence of successive disintegrations, all members of the disintegration series have mass numbers given by $4n+m$, with n variable.

[e] Widdowson and Russell, *Phil. Mag*. 48, 293, 1924.

[f] Recently, by bombarding bismuth with deuterons, radium E has been produced, with properties entirely those of the naturally occurring product: see Livingood, *Phys. Rev*. 49, 876; *ibid*. 50, 425, 1936.

§ 27. *Positive electrons of secondary origin.* The positive electron was discovered first[a] as one component of the penetrating radiation at sea-level. A little later[b] it was found to be produced by the action of high energy γ radiation upon matter and then[c] (much more rarely) as a result of close collisions of negative electrons and nuclei.[d] Generally, in the two latter cases, the appearance of each positron was found to be associated with the simultaneous appearance of a negative electron of comparable speed. Energy is said to undergo "materialisation", with conservation of electric charge, in these transformations. Because the numbers of positive and negative electrons in the penetrating radiation at sea-level are approximately equal it is generally believed that processes of materialisation are to a large extent responsible for the secondary electrons there observed. With these questions, however, we are not concerned in this book. Mention of them is made only as a preface to quite different considerations. There has been definite evidence for some time[e] that sources of radon, radiothorium and thorium active deposit (enclosed in thin walled glass tubes) emit positive as well as negative electrons (up to about 1 per cent. of the total β particle emission), and a general indication that many of these positrons arise in the radioactive material itself. In view of the fact that nuclear disintegration with the emission of positrons is a possible occurrence, it is important to decide whether the phenomena described are of primary or secondary origin.

[a] Anderson, *Science*, 76, 238, 1932; Blackett and Occhialini, *Proc. Roy. Soc.* 139, 699, 1933.

[b] Anderson, *Science*, 77, 432, 1933; Chadwick, Blackett and Occhialini, *Proc. Roy. Soc.* 144, 235, 1934; Crane, Delsasso, Fowler and Lauritsen, *Phys. Rev.* 46, 531, 1934.

[c] Anderson, Millikan, Neddermeyer and Pickering, *Phys. Rev.* 45, 352, 1934.

[d] The γ ray and electron effects have generally been described as "nuclear" processes—merely because the neighbourhood of an atomic nucleus is the only place where electric fields are of sufficient intensity to give appreciable probability of occurrence to the process in question: see Oppenheimer and Plessett, *Phys. Rev.* 44, 53, 1933; Bethe and Heitler, *Proc. Roy. Soc.* 146, 83, 1934; Jaeger and Hulme, *ibid.* 153, 443, 1936.

[e] Thibaud, *Comptes rendus*, 197, 915, 1933; Chadwick, Blackett and Occhialini, *loc. cit.* Skobelzyn and Stepanowa, *Nature*, 133, 565, 646, 1934.

It may be said at once that they are of a secondary nature. First, Alichanow and Kosodaew[a] examined in greater detail the positrons from a radon source and then the radiation from a bare source of thorium active deposit.[b] Energy distribution curves are given in Figs. 12 and 13. Ascribing the emission to the C and C″ bodies in each case, the total intensity was found to be of the order of 3×10^{-4} positron per

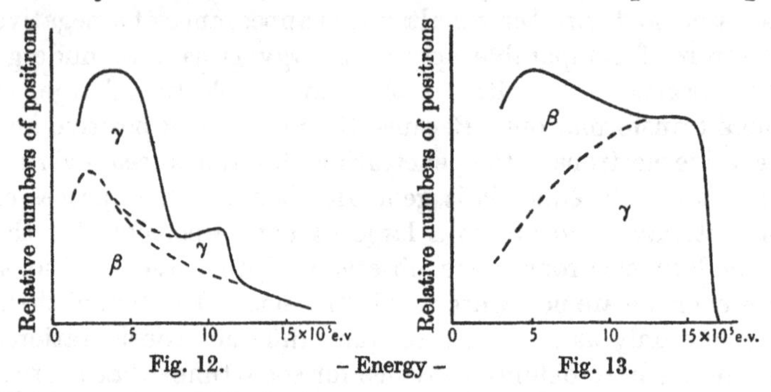

Fig. 12. – Energy – Fig. 13.

disintegration. This is, in part, attributed to "internal conversion" of γ rays with the production of pairs[c] and, in part, to a similar internal conversion effect for the primary β particles. The suggestion appears qualitatively and quantitatively sound: intensities are in good accord with the calculations[d] and the upper limits of the partial energy distributions, indicated by the dotted lines in the diagrams, are generally about 1 million electron volts[e] lower than the γ ray energies and the values of E_0 for the primary β particle spectra. This is an interesting result and it emphasises the importance of internal conversion effects in general. Of these

[a] Alichanow and Kosodaew, *Z. Physik*, 90, 249, 1934; Alichanow, Alichanian and Kosodaew, *Nature*, 136, 719, 1935.

[b] Alichanow, Alichanian and Kosodaew, *Nature*, 136, 475, 1935; *J. Physique*, 7, 163, 1936.

[c] This was the suggestion of Curie and Joliot concerning the positive electrons found to be emitted by beryllium under α particle bombardment (p. 126). High energy γ rays are known to result from the disintegrations produced.

[d] Nedelsky and Oppenheimer, *Phys. Rev.* 44, 948, 1933; *ibid.* 45, 136, 283, 1934; Jaeger and Hulme, *Proc. Roy. Soc.* 148, 708, 1935; Møller, *Nature*, 137, 314, 1936.

[e] Energy of this amount is absorbed in the creation of the mass of the two electrons.

the best known, it may be remarked, is responsible for the greater part of our detailed knowledge of the γ ray emission from radioactive substances, which is discussed in chap. IX.

§ 28. *Theories of β particle disintegration.* Until quite recently there was no satisfactory theory of β particle disintegration; the extremely successful application of the principles of wave mechanics to the explanation of the main features of the α particle transformation (§ 24) had no counterpart in the other case. Difficulties were chiefly those of describing the behaviour of nuclear electrons on any admissible assumptions. Reference to these difficulties has already been made (§ 12). With the discoveries of the neutron and positive electron, however, a new outlook became possible; as a result of this two distinct theories have been put forward and have achieved some measure of success. The earlier theory is due to Beck,[a] the later—and more widely accepted—theory was developed primarily by Fermi.[b] Both are based on the same general interpretation of the experimental material and have certain formal similarities, but, fundamentally, they are quite distinct. We shall deal first with the theory of Fermi.

In this theory the first hypothesis is as follows: it is assumed that the emission of a positive or negative electron from any nucleus is the result of the transformation of a proton into a neutron (or of a neutron into a proton) within the nuclear structure. This is in line with the general supposition that only neutrons and protons have any individual existence as structural units. The transformation here postulated has been likened by Fermi to the emission (or absorption) of light by an atom,[c] neutron and proton being regarded as different inner quantum states of a single "fundamental heavy particle". On this view, the probability of the emission of electron or positron is merely the probability of the quantum jump between the appropriate states of this particle. This automatically provides the reason why the

[a] Beck, *Z. Physik*, 83, 498, 1933. [b] Fermi, *Z. Physik*, 88, 161, 1934.

[c] Some years previously this analogy had similarly been employed in respect of β disintegration: cf. Ambarzumian and Iwanenko, *Comptes rendus*, 190, 582, 1930.

atomic number of the disintegrating nucleus is so little involved in determining the probability in question.

The second assumption—which, however, has a marked influence on the calculation of these probabilities—is really concerned with the principle of the conservation of energy. It is based upon the conclusion that the limiting energy in the β particle spectrum is to be interpreted as the unique energy of disintegration for the change which is under consideration—and it derives directly from a suggestion made by Pauli in 1931.[a] Pauli assumed that two particles, not one, are emitted in every act of β disintegration, with the available energy shared between them. The second of these particles, the "neutrino", was supposed to be uncharged and to have a very small mass. In Fermi's theory it is also regarded as characterised by a spin quantum number, I, equal to $\frac{1}{2}$.[b] It is assumed then, that the quantum jump of the fundamental heavy particle results in the emission of an electron (positive or negative) and a neutrino, for which the sum of the kinetic energies is constant for nuclei of any one species.[c]

Clearly, the theory of Fermi can only be completely substantiated by an independent proof of the emission of the neutrino. This point will be discussed in greater detail later; for the present it is worth while examining the success which the theory achieves, once its major premiss is granted. It will be discovered that, even then, success is not achieved without additional particular assumptions being made.

The aim of any theory of β disintegration must be to make possible calculations concerning both the energy distribution function $F(E)$ in respect of the disintegration electrons and also the experimental transformation constant λ, as a function of the disintegration energy (E_0)—and other parameters. For Fermi's theory to do this, particular assumptions must be

[a] See Carlson and Oppenheimer, *Phys. Rev.* 38, 1787, 1931.

[b] It has become increasingly clear as data accumulate (cf. Table 7) that the change in nuclear angular momentum on β disintegration must always be represented by an integral change in I. Since $I=\frac{1}{2}$ for the electron, this assumption ($I=\frac{1}{2}$ for the neutrino) is the simplest under which angular momentum may be conserved.

[c] Essentially the same idea was worked out somewhat earlier by Perrin (*Comptes rendus*, 197, 1625, 1933), though without similar elaboration.

made which involve the mass of the neutrino, the details of the interaction between the heavy particles and the "light particle field" and the effective radius of the nucleus. It is to the credit of the theory that the assumption of zero rest mass for the new particle is in best accord with facts of experiment.[a] As regards the other assumptions, however, the position at first was less satisfactory. The empirical $\lambda - E_0$ relation could only be obtained by employing such numerical constants in the interaction function that the derived mutual energy of neutron and proton was quite useless for calculations in respect of binding energy (p. 65)—or the scattering of neutrons by nuclei;[b] moreover, an uncomfortably large nuclear radius had to be assumed to explain why the distribution function in the case of positron emission was so little different from that appropriate to the emission of negative electrons.[c] Reference has already been made (p. 69) to certain attempts to remove these discrepancies; for further discussion, however, the reader should consult the work of Gamow.[d]

Here we must revert, for a moment, to the $\lambda - E_0$ relation of Sargent (Fig. 10). Hitherto no mention has been made of the two curves which account for the great majority of experimental points in the diagram. It is again to the credit of Fermi's theory that a completely natural suggestion can be made concerning them. It is suggested that the upper curve belongs to disintegrations which take place without change of spin, the lower curve to those disintegrations for which $\Delta I = 1$.[e] Without any particular assumption, the ratio

[a] Perrin, *loc. cit.*; Henderson (W. J.), *Proc. Camb. Phil. Soc.* 31, 285, 1935.

[b] Nordsieck, *Phys. Rev.* 46, 234, 1934. The sense of the discrepancy may be stated as follows: Assuming the force function constants otherwise required, β particle disintegration probabilities might be expected to be much greater than, in fact, they are found to be.

[c] Perrin, *Comptes rendus*, 198, 2086, 1934.

[d] Gamow, *Structure of Atomic Nuclei and Nuclear Transformations*, 1936, chap. VII.

[e] Disintegrations for which $\Delta I = 0$, 1 are sometimes loosely referred to as "permitted" and "forbidden" transitions, respectively. The disintegrations of potassium and rubidium are probably more strongly "forbidden", by the greater change of spin which is involved. For a more recent discussion of the selection rule, see Gamow and Teller, *Phys. Rev.* 49, 895, 1936.

of the respective transformation probabilities, for a given energy of disintegration, may be shown to be about 100 : 1. This is not very different from the mean ratio suggested by the curves.[a]

Finally, direct experimental evidence concerning the emission of neutrinos must receive brief attention. So far such evidence is almost entirely negative; no definite proof has been obtained of any ionisation to be attributed to such particles in their passage through matter. There are, however, circumstances in which no ionisation would be expected. If the neutrino possesses mechanical, but no magnetic, moment then its only interaction will be with an atomic nucleus, in a process which is the direct reverse of the act of emission.[b] Any experimental evidence must rather be regarded, therefore, as setting an upper limit to the magnetic moment of the hypothetical particle; it cannot be used to disprove its existence. The relevant theory has been developed by Carlson and Oppenheimer[c] and by Bethe;[d] the experiments are due to Chadwick and Lea[e] and to Nahmias.[f] The final result may be stated as follows: If neutrinos are emitted as Fermi's theory predicts, then a neutrino of average energy from radium B + C produces less than 1 pair of ions in 3×10^5 km. of air at N.T.P. and has a magnetic moment certainly less than 2×10^{-4} Bohr magneton. For lack of any evidence for the particle in its interaction with matter it becomes necessary to look for an effect on the emitting nucleus itself. It is a question whether the recoil of the nucleus, following β disintegration, corresponds to the momentum gained from the β particle alone, or to the resultant momentum of β particle and neutrino.[g] It will be realised that, both experimentally and theoretically, considerable difficulties are involved. In any case, the energy of β recoil

[a] See, however, Sitte, *Int. Conf. Phys.* 1, 68, 1935.
[b] Bethe and Peierls, *Nature*, 133, 532, 1934.
[c] Carlson and Oppenheimer, *Phys. Rev.* 41, 763, 1932.
[d] Bethe, *Proc. Camb. Phil. Soc.* 31, 108, 1935.
[e] Chadwick and Lea, *Proc. Camb. Phil. Soc.* 30, 59, 1934.
[f] Nahmias, *Proc. Camb. Phil. Soc.* 31, 99, 1935.
[g] See Leipunski, *Proc. Camb. Phil. Soc.* 32, 301, 1936.

from a heavy radioelement is of the order of 1 electron volt—which may easily be masked by spurious surface effects.

The theory of Fermi, as we have seen, is based on the assumption that electrons do not already exist in the nucleus. From the point of view of Beck this is a very dangerous and illogical assumption. Inability to treat theoretically the binding of electrons in nuclei is, for him, evidence for the present inadequacy of the theory, not for the absence of the electrons. He does not, therefore, propose a detailed theory, but a tentative method of description which stresses the points of failure of accepted ideas. He regards nuclei as built up of α particles, not more than three protons and the requisite number of "free charges" (cf. p. 57)—entities which may be regarded as electrons which, within the nucleus, retain only their charge and symmetry properties, but no purely mechanical attributes. In β disintegration there is the materialisation of energy within an unstable nucleus and this occurs, as in other known cases, with the creation of a pair of electrons, positive and negative. Only one of these electrons escapes from the nucleus.[a] The capture of the other takes place, in a manner which present theory is unable to describe, so that only the charge of the particle is retained:[b] its energy is lost to observation. Thus, the features of β disintegration which on Fermi's theory are described in terms of the emission of an unobserved particle, the neutrino, are here represented by this process of capture of the second electron with the loss of its purely mechanical attributes. No effort is made to save the conservation laws of macroscopic physics. Yet considerable progress, in fact, has been made with this rather vague theory—and, because of its close formal similarity with the more orthodox theory of Fermi (on both theories disintegration is treated as a double process, involving two light particles), the results which have been obtained are not very different, either.[c]

[a] Whether positive or negative electron escapes is decided by the type of instability.

[b] A suggestion of this type was also made by Saha and Kothari, *Nature*, 132, 747, 1933; *ibid.* 133, 99, 1934.

[c] Cf. Beck, *Int. Conf. Phys.* 1, 31, 1935.

CHAPTER IX

EMISSION OF QUANTA

§ 29. *Experimental facts.* Information concerning the emission of quanta in radioactive disintegration would be regarded as complete—from the purely experimental point of view—if the quantum energies, the intensities (quanta per disintegrating atom) and the time intervals involved in the emission from each type of atom were known. Direct information concerning the time intervals is almost entirely lacking. It is evident that, in general, the interval of time between the escape of the disintegration particle and the emission of the quantum is longer than the time of relaxation of the more tightly bound atomic electrons (§ 22), but all attempts at obtaining positive indications of a finite lapse of time have so far resulted in failure.[a] Such experiments, however, are inconclusive for times less than about 10^{-6} sec. From quite indirect evidence[b] it is probable that in many cases γ ray transition probabilities of the order of 10^{11} sec.$^{-1}$ (half-value periods $\sim 10^{-11}$ sec.) must be assumed. The comparative smallness of these probabilities is a somewhat unexpected result.

Quantum energies may be deduced in a number of ways,[c] by far the most important of these being the method of the natural β ray spectrum. The observational material consists of measurements on the magnetic line-spectrum of the electrons arising in the internal (photoelectric) conversion of the γ rays in the atom of origin. Obviously, values for the ionisation energies of the atomic levels must be taken over from (interpolated) X-ray data, following a decision regarding the effective atomic number of the atom at the moment of

[a] Joliot, *Comptes rendus*, 191, 132, 1930; Wright, *Proc. Camb. Phil. Soc.* 28, 128, 1932.

[b] Cf. Gamow, *Structure of Atomic Nuclei and Nuclear Transformations*, 1936, chap. VI.

[c] Cf. Rutherford, Chadwick and Ellis, *Radiations from Radioactive Substances*, 1930, chap. XII, particularly §§ 83, 86.

absorption (this is the decision regarding priority of particle or quantum, already discussed). If the uniform field in the magnetic spectrograph is H, the energy of the photoelectrons focused at a distance 2ρ from the source (energy expressed as V electron volts) is given by the relation[a]

$$V=\frac{c}{10^8}\left\{\left[(H\rho)^2+\left(\frac{m_0c}{e}\right)^2\right]^{\frac{1}{2}}-\frac{m_0c}{e}\right\} \quad \ldots\ldots(29).$$

Here c is the velocity of light in cm. per sec. and e/m_0 the specific charge of the electron at rest (electromagnetic units per gm.). Then the quantum energy, $h\nu$ (ergs), may be calculated in terms of V and the appropriate ionisation energy W (electron volts) by means of the equation

$$h\nu=10^8\,(V+W)\,e \quad \ldots\ldots(30).$$

As before, the electronic charge, e, is expressed in electromagnetic units.

The most direct method of investigating the distribution of intensity in the γ ray spectrum of an element involves a combination of the analysis of the energies and directions of projection of the recoil electrons produced in an expansion chamber with a measurement of the total heat evolved in the absorption of a known (large) fraction of the γ radiation in a suitable calorimeter.[b] The former analysis provides relative intensities over different portions of the spectrum (the method is not one of high resolution), whilst the latter determination gives the total intensity of the radiation. The validity of the general scattering formulae and of conservation of energy and momentum in individual processes is assumed.[c] More detailed, but less direct, information concerning intensities may be deduced from the natural β ray spectra themselves. This, however, involves a theory of internal conversion coefficients. Originally it was hoped that the direct determina-

[a] This equation is given because the necessity may again arise, as it has done in the past, of recalculating γ ray data because of a change in the accepted value of e/m_0. At present $1{\cdot}760\times10^7$ e.m.u./gm. is generally adopted, although it appears that even this value is too high.

[b] Cf. Rutherford, Chadwick and Ellis, *loc. cit.* chap. XVI.

[c] Detailed conservation has recently been questioned as a result of the experiments of Shankland (*Phys. Rev.* 49, 8, 1936), but further investigation has failed to confirm this negative result.

tion of these coefficients for the most intense γ rays, by the comparison of intensities of corresponding lines in the natural and "excited"[a] β ray spectra, might provide a sufficient basis for interpolation in other cases, but this proved to be quite impracticable.[b] Experimentally determined internal conversion coefficients did not vary monotonically with the energy, even for a single radioelement. The relevant theory has been developed by Hulme,[c] Taylor and Mott,[d] Fisk,[e] Fisk and Taylor[f] and Taylor.[g] It appears that the value of the internal conversion coefficient depends very markedly upon the type of nuclear transition in which the γ ray quantum is produced. For electric quadripole transitions (in which the nuclear azimuthal quantum number changes by 0 or 2) it is about three times greater than for dipole transitions (in which the change of spin is one unit)—although this result is considerably modified (and much larger coefficients are obtained) if radiation fields involving magnetic multipole components, also, are employed to describe the transitions. This theory of internal conversion provided an immediate qualitative explanation of the apparently anomalous results of the early experiments: obviously different types of nuclear transition were involved in the emission of radiation.[h] Always neglecting any complication by "magnetic components", it became possible to classify the more intense radiations as of one or other transition-type (experimental values were of the predicted order)—and, for the others, to postulate alternative values for the coefficient of internal conversion depending upon the radiation-type assumed. The

[a] The spectrum of the photoelectrons produced in a thin sheet of absorbing material (placed round the source in a magnetic spectrograph) is referred to as the "excited" spectrum. The law of external photoelectric absorption is fairly well known, but considerations of intensity make the method impossible except with the strongest γ rays.

[b] Ellis and Aston (G. H.), *Proc. Roy. Soc.* 129, 180, 1930.

[c] Hulme, *Proc. Roy. Soc.* 138, 643, 1932.

[d] Taylor (H. M.) and Mott, *ibid.* 138, 665, 1932.

[e] Fisk, *ibid.* 143, 674, 1934.

[f] Fisk and Taylor, *ibid.* 146, 178, 1934.

[g] Taylor, *Proc. Camb. Phil. Soc.* 31, 407, 1935; *ibid.* 32, 291, 1936. See, also, Hulme, Mott, Oppenheimer (F.) and Taylor, *Proc. Roy. Soc.* 155, 315, 1936.

[h] Practically speaking, only dipole transitions occur in the outer atom.

decision between such alternatives was then made—in cases in which this decision proved possible—by methods to be described in the next section. Thus, finally, γ ray intensities have been obtained for many spectra.

Before discussing this matter further, however, it is necessary to stop to review the main features of the complete data at present available. This information is so extensive that no excuse is necessary for the failure to tabulate it in full here; it may be found in the standard works and in recent papers.[a] We wish, first of all, to compare the general features of the γ ray spectra of α and β active elements of the main series. There are clearly-marked differences between them. There is, however, this similarity, also; in both cases examples occur of disintegrations which are not followed by the emission of γ radiation—or followed by radiation of extremely small intensity, only. Such are the β particle disintegrations of uranium X_2, radium E and thorium C and several α particle disintegrations—more particularly of elements belonging to the uranium and thorium series. The differences chiefly concern the extent of the γ ray spectra in the two cases: with β active elements it is frequently found (cf. p. 123) that γ rays of high energy are emitted (low energy γ rays are then always emitted, as well), but with α active elements the γ ray of highest energy so far discovered has less than one-eighth of the energy of the α particle which it follows. With all α active elements the spectrum of γ rays is confined to the region below $6{\cdot}5 . 10^5$ electron volts; in this region, however, it resembles in all essential particulars the spectrum of the γ radiation from a representative β active element. We shall later bring these facts into relation with another point of difference between α and β disintegrations, which, since it has not previously been referred to, may be mentioned here. Confining attention entirely to elements for which both decay constant and disintegration energy are reasonably well known, and, for β active bodies, to those belonging to the same branch of the Sargent curve (Fig. 10), we have the results, given on p. 142, for the extremes in λ and E.

[a] E.g. Ellis, *Proc. Roy. Soc.* 138, 318, 1932; *ibid.* 143, 350, 1934.

The enormous disparity between the rates of change of decay constant with energy of disintegration will be immediately obvious.

α-active		β-active	
λ: $1{\cdot}7\times10^{-18}$ sec.$^{-1}$,	E: $4{\cdot}2\times10^{6}$ e.v.	λ: $1{\cdot}6\times10^{-9}$ sec.$^{-1}$,	E: $0{\cdot}22\times10^{6}$ e.v.
$3{\cdot}5\times10^{3}$ sec.$^{-1}$,	$7{\cdot}8\times10^{6}$ e.v.	$5{\cdot}9\times10^{-4}$ sec.$^{-1}$,	$3{\cdot}15\times10^{6}$ e.v.

The second general consideration refers to the "artificially" produced radioelements of short life. The emission of γ radiation in such cases was first demonstrated by Fleischmann,[a] by the ionisation (absorption) method. This was for the negative electron active species formed by bombarding iron, aluminium and silicon with neutrons. Soon afterwards many other cases of a similar nature were discovered.[b] Because of the generally small intensity of "artificially radioactive" sources (and the smallness of the atomic numbers involved) it is not as yet possible to employ the method of the magnetic spectrograph for the more exact determination of energies; probably the most accurate method, therefore, is that of recoil electrons in the expansion chamber. The use of this method has been discussed by Richardson and Kurie[c] and preliminary results obtained by means of it have already been reported.[d] At present, however, data of this type are neither sufficiently accurate nor extensive to make further discussion of them profitable. Suffice to say that until now no certain evidence of the emission of γ radiation after positron disintegration has been set in evidence. It would indeed be remarkable if this state of affairs were to persist for long.

§ 30. *Nuclear energy levels.* It was obvious at an early stage that[e] the energies of the γ rays from a number of radioactive bodies afforded several examples of the applicability of the combination principle of Ritz. It was plausible to assume, therefore, that further research would establish unambigu-

[a] Fleischmann, *Naturwiss.* 22, 434, 1934.

[b] Amaldi and others, *La Ricerca Scientifica*, 5 (1), 452, 1934.

[c] Richardson and Kurie, *Phys. Rev.* 49, 209, 1936.

[d] Richardson, *ibid.* 49, 203, 1936; Cork, Richardson and Kurie, *ibid.* 49, 208, 1936.

[e] Ellis, *Proc. Roy. Soc.* 101, 1, 1922.

ously systems of nuclear energy levels capable of describing in detail the nature of the γ radiation from the nuclei in question. Only recently, however, has any approach to such certainty been attained. The trouble has been, in great part, the lack of precision in the measurements of energy. Compared with the data of optical or X-ray spectroscopy the relative energies of the γ rays, carrying possible errors of a fraction of one per cent., are extremely crude. Further difficulties, however, have until recently been caused by the complete lack of additional criteria to apply to the level systems proposed. There was no lack of suggestions, rather was it impossible to judge between a number of schemes, all of which were equally plausible from energy considerations alone. The present improvement in the situation is to a great extent due to the possibility of applying the criteria which were previously lacking. These criteria depend upon a detailed consideration of intensities. By the time that intensities have been assigned to the various components of the γ ray spectrum it has also been decided which radiations are dipole and which quadripole (p. 140). As previously mentioned, this corresponds to a decision regarding the possible changes of nuclear spin in each act of emission. Clearly, the successful level system is one in which spin quantum numbers may be assigned to the levels in such a way that the correct radiation type is predicted for each component of the spectrum. Again, before intensities are known, there is nothing to show which levels in any given scheme are levels of high excitation and which of low. When intensities have been assigned this ambiguity is resolved. With the successful level system it must necessarily be resolved in the same sense when each level is separately considered in turn. To do this we investigate the total intensity of transitions "to" each level and the intensity "away from" that level. These directions ("to" and "from") have been correctly chosen when the latter intensity is not less than the former, transitions from the level being those which result in the emission of radiation. The excess of transitions from any level over those in the opposite direction (to that level) is obviously the extent of excitation of the level as a result of the previous particle disintegration. A final criterion is, clearly, that the sum of the primary excitations of the levels above the ground level shall not be greater than unity. When this sum is less than unity the balance obviously represents the fraction of cases in which the particle disintegration takes place without excitation of the residual nucleus. As an instance of a successful level scheme and, also, as providing a particular example with the help of which to study further the above generalisations, Fig. 14 gives the

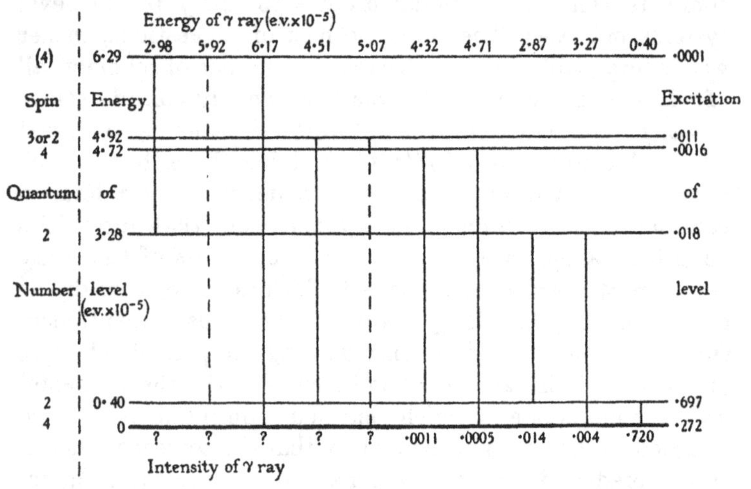

Fig. 14.

complete scheme of levels and transitions for the nucleus of thorium C″ which is left excited to emit γ radiation following the (α particle) disintegration of thorium C.[a] In constructing this diagram the values given for the intensities of the various radiations have been slightly idealised for the sake of numerical neatness. The diagram itself will be further referred to in subsequent discussions (p. 147).

One question concerning nuclear levels appears somewhat to have lost importance in recent years. Previously it was frequently enquired what type of particle should be regarded

[a] Ellis, *Int. Conf. Phys.* 1, 43, 1935; Rosenblum, *Comptes rendus*, 202, 943, 1936.

as making the transition from one level to another in the act of emission. In a sense there were difficulties with either assumption—that electron or α particle levels were involved. As we have seen, the excitation of the emitting system may be occasioned by the emission from the nucleus either of an α particle or a β particle. Then, at one stage, it was considered that the matter had been satisfactorily settled: the levels in question were thought to be α particle levels. But this investigation of Kuhn[a] (p. 201) showed that the high degree of homogeneity of the various components of γ radiation merely required that the specific charge of the particle making the transition be not very different from that of the nucleus as a whole. The electron certainly was excluded, but the whole nucleus remained as a possible vehicle for the mechanism of emission. To ascribe such an origin to the radiation—and this significance to the system of energy levels—is entirely in line with the present outlook of theoretical nuclear physics.

§ 31. *Evidence for nuclear excitation in the analysis of the primary corpuscular radiation.* Adopting a general viewpoint, based upon the conservation of energy in individual processes, one would expect to find other evidence for the excitation of nuclei than the emission of γ rays. If the total amount of energy set free, say from uranium to lead, is always the same, then in those cases in which quanta of γ radiation are emitted during the course of the transformations, some particles must be ejected with correspondingly less than the normal energy.[b] Moreover, since the nuclei with which we are concerned are unstable against particle disintegration when in the ground state, there is always the possibility of the ejection of particles (rather than radiation) from nuclei which are temporarily excited. Such particles would be expected to possess more than the normal amount of energy. Actually, portions of this evidence had been obtained before its true bearing was appreciated. It has already been discussed—purely empirically—in § 23.

[a] Kuhn, *Z. Physik*, 44, 32, 1927; *Phil. Mag.* 8, 625, 1929.

[b] Here it is tacitly regarded as established that γ rays with fractional intensities are involved. This is almost always the case.

We shall consider first the fact of the emission of α particles of long range. This is interpreted now[a] as an ejection of particles from excited nuclei. Long-range α particles have been found with certainty only with radium C′ and thorium C′—and most abundantly, with respect to the main group, in the latter case. Again, these two bodies, radium C′ and thorium C′, are the α active elements of shortest life, and, of the two, thorium C′ has the shorter period. These facts are not accidentally related; they fit naturally into the scheme of explanation which has been proposed. The intensity of a group of long-range particles is obviously determined by the intensity of excitation of the corresponding state of the nucleus and the ratio, $\lambda_\alpha{}^r/\lambda_\gamma{}^r$, of the transformation constants for particle emission and radiative transformation from the state in question. Clearly, this ratio has the best chance of being appreciable when $\lambda_\alpha{}^0$, the disintegration constant for α emission from the ground state, is already large. As indicated above, the last condition is best fulfilled with radium C′ and thorium C′. To provide quantitative verification for our hypothesis we have to enquire whether excitation energies for the nuclei of radium C′ and thorium C′, obtained by subtracting the normal energy of α disintegration in each case from the various "abnormal" values listed in Table 10, are consistent with the evidence concerning the γ rays associated with the antecedent disintegrations RaC.C′ and ThC.C′. Until recently no evidence for γ radiation associated with the latter transformation had been obtained, but Ellis[b] has now found just such (weak) natural β ray lines as are required to supply this deficiency: in this case the evidence is clear-cut, and, in respect of energies at least, entirely satisfactory. With radium C′ there is no dearth of data regarding quantum radiations, roughly sixty different frequencies being represented in the complete γ ray spectrum for the transformation RaC.C′. Oppenheimer[c] has investigated the particular problem of accommodating these radiations in a system of

[a] Feather, *Phys. Rev.* 34, 1558, 1929.

[b] Ellis, *Proc. Roy. Soc.* 143, 350, 1934.

[c] See Ellis, *Int. Conf. Phys.* 1, 43, 1935; also for similar investigations, Oppenheimer (F.), *Proc. Camb. Phil. Soc.* 32, 328, 1936.

levels for which the long-range α particle levels provide the original framework. Again, so far as energy values are concerned, the fit is altogether satisfactory. However, the problem will not be completely solved until a similar conclusion has been reached concerning intensities also. Here, whilst there is no reason to anticipate ultimate failure, at present success is not fully achieved—and the reader is referred to original sources for further information on the subject. We may sum up by repeating the initial assertion: long-range α particles are those which are emitted from excited nuclei as infrequent alternatives to the emission of radiation.[a] For that reason, if the experiment could be performed, it would certainly be found that the long-range and the normal α particle activities of a pure source of C′ product did not decay together: initially chiefly long-range particles would be emitted, then, for the rest of the time (by far the greater portion), almost wholly α particles of the normal energy (cf. p. 116).

The suggestion that the fine structure of the normal α particle emission, found with some elements, should be regarded as evidence for the excitation of the subsequent unstable nucleus was first made by Gamow.[b] This suggestion has turned out eminently correct. Fig. 14 has already been described as the level system for the γ rays arising in the transition ThC.C″. Primarily, however, these levels represent the energy differences between the various fine structure α particle groups of thorium C. In so far as energies are concerned it appears that the γ rays which follow[c] the emission of α particles from this product fit excellently into the set of levels so constructed: when an α particle of less than the normal energy is emitted, the nucleus is left to get rid of its excess energy in the form of radiation. The question of intensities, however, is still to be considered. If the present

[a] In some cases the direct emission of radiation is a relatively improbable occurrence and the important alternative is the de-excitation of the nuclear state by the "mechanical" intervention of an extra-nuclear electron. This is the reason for the great intensity of the third long range α group from RaC′. The corresponding radiative transition is "forbidden".

[b] Gamow, *Nature*, 126, 397, 1930. [c] Ellis, *Proc. Roy. Soc.* 136, 396, 1932.

view is correct, then the relative intensities of the fine structure groups give at once the amounts of excitation of the various levels of the residual nucleus. We have to understand these intensities—and to see whether they, too, are consistent with the γ ray evidence. As concerns the latter point, in the construction of Fig. 14 (using, as stated, slightly idealised values) it has already been assumed that there is no inconsistency—this is the conclusion which has been reached as a result of detailed investigations[a]—but it is worth while pursuing the former question a little more fully.

Qualitatively the type of explanation is perfectly evident—and here, again, we shall leave the problem at the qualitative stage. We have to consider the spins of the original nucleus and of the various states, normal and excited, of the subsequent product (cf. § 24). It may happen that the nuclear spins are the same for the normal states of these two nuclei. Then the transition between these states, obviously involving a greater release of energy than any other, is also very much more probable than the transition to any higher state. "Fine structure" is unlikely to be observed. On the other hand, the normal state of the initial nucleus may have the same spin as a possible excited state of the product. Then, if the energy of excitation of this state is not too high (for disintegration probabilities decrease rapidly as the effective disintegration energy becomes smaller), the transition to it may be of intensity comparable with that of the more energetic change which results in the final nucleus in the ground state (of different spin). As a result, "fine structure" will be observed. In Fig. 14 spin quantum numbers have been chosen (with reference to that of the normal thorium C nucleus, assumed to be 2) to be in qualitative agreement with what might be expected from the intensities of the corresponding α particle groups. In particular, with this set of spins,[b] it will be evident why the partial probability of disintegration to the first excited state of thorium C″ is even greater than the probability of the "normal" disintegration: the energies involved

[a] Ellis and Mott, *Proc. Roy. Soc.* 139, 369, 1933.

[b] Spin-*differences* are fixed chiefly from γ ray considerations (p. 140).

are not very different and the spin changes are, respectively, zero and two quantum units of spin.

Hitherto, only α particle data have been discussed as providing evidence for the excitation of nuclei. On the basis of the same views regarding energy conservation, however, similar evidence must also be provided by any complete set of data concerning β particle disintegrations. It is generally agreed that it is so provided, but, because of the nature of the case, it is much more difficult to disentangle. Instead of a discrete fine structure in the α particle energy spectrum—indicating possible excitation of the subsequent nucleus—we have to look for the superposition of partial continuous β particle spectra with different upper limits. According to our basic ideas, it is the spectrum of these limiting energies which gives the energy levels of the next product, and the integrated areas of the partial continuous distributions which correspond to the intensities of excitation. Ellis and Mott[a] made the first fairly successful attempt to apply these ideas, working backwards from the levels and excitations, obtained from other evidence for the γ rays of RaC.C′, to the composite form of the continuous β particle spectrum for radium C.[b] It is now known that the matter is not so simple as they supposed,[c] but there is equally no doubt that their fundamental hypothesis is thoroughly sound.[d] In what follows, therefore, we may understand "disintegration" to refer either to α particle or to β particle transformations—and we shall attempt, as originally intended (p. 112), to put a modern

[a] Ellis and Mott, *Proc. Roy. Soc.* 141, 502, 1933.

[b] By direct coincidence experiments, v. Baeyer (*Z. Physik*, 95, 417, 1935) has recently shown that hard γ rays are chiefly emitted from those atoms which have just previously ejected an electron of sub-normal energy. This investigation, also, refers to the transformation RaC.C′.

[c] Cf. Gamow, *Structure of Atomic Nuclei and Nuclear Transformations*, 1936, chap. VIII.

[d] The result (p. 123) that γ ray energies are usually less than the maximum energy of the β particles which precede them is clearly in line with this hypothesis. Ellis and Mott explained the contrary result with ThC″ by assuming that the production of an unexcited nucleus (Pb^{208}) is so improbable that the corresponding high energy β particles entirely escape detection. We have already (p. 125) made successful use of the suggestion that γ ray energy, to the extent of $3{\cdot}2 \times 10^6$ e.v., is always emitted along with the β particles forming the observed distribution.

interpretation upon the concept of the disintegration series.

Essentially we have to do with the sequence of changes by which the nucleus of an atom of, say, uranium I is finally transformed into a lead nucleus, Pb^{206}. This sequence, which comprises particle disintegrations and radiative transformations, is by no means invariable, although it involves the release of a total amount of energy which is always the same. Nuclei in the ground state and excited nuclei are concerned in the changes, and, at each stage, any transformation which is energetically possible must be thought of as possessing finite probability.[a] Sometimes both α particle and β particle disintegrations are energetically possible (α–β branching occurs[b]) but usually one mode only is found. Alternatively, disintegrations resulting in excited nuclei (and disintegration particles of less than the normal energy) may also possess partial probabilities comparable with those describing the normal mode. This is more frequently so—and the range of energies of excitation is generally greater—when β disintegration is in question (cf. p. 141). Excitation occurs mainly when the normal states of the initial and final nuclei differ in angular momentum, and the greater range[c] of excitation by β disintegration is clearly connected with the slower variation of (partial) disintegration probability with particle energy in this case (cf. p. 142). No penetrating γ rays follow α particle changes merely because of the high improbability of the previous emission of an α particle of correspondingly small energy. Finally, when disintegration, rather than radiative transition, takes place from an excited state, in actual experience it is an α particle change which is involved. The same explanation applies: within the range of energies in question only α particle changes (and then only the most energetic of

[a] This point of view was clearly expressed first by Gurney and Condon, *Phys. Rev.* 33, 127, 1929.

[b] Also, one example of electron-positron branching has recently been reported: see van Voorhis, *Phys. Rev.* 49, 876, 1936.

[c] The greater *likelihood* of excitation following β disintegration must imply that a change of nuclear spin is more probable in this process than in α emission.

such changes) possess probabilities in any way comparable with those which belong to the radiative processes of the nucleus. It is just because these probabilities are so enormously greater than the probabilities of α or β disintegration in general that the original simple concept of the disintegration series assumed the importance which it did.

PART FOUR

CHAPTER X

TRANSFORMATIONS PRODUCED BY α PARTICLES

§ 32. *General types of artificially produced nuclear transformation.* Nuclear transformations may be induced artificially[a] by subjecting matter to the bombardment of fast-moving atomic projectiles or to radiation of sufficiently high frequency. Then, in spite of the very small target area which an atomic nucleus presents to such a bombardment, it is merely necessary to employ intense enough sources of particles or radiation, and to adopt methods of detecting individually the products which result from the transformations, to be in a position to study the very varied effects which occur. Already (§§ 2, 5) we have discussed the problem of the detection of individual particles and have examined the methods employed for the production of the intense sources required for the bombardment; here, before treating of the effects more systematically and in detail (effects due to particles in chaps. X, XI and XII; those produced by radiation in chap. XIII), it will be well to classify briefly the types of transformation which may be expected[b] and to indicate the broad features of the phenomena to which it is most important that attention should be given. Moreover, in order to simplify the present discussion, the question of transformations produced by radiation will not be included at this stage (see § 44).

[a] The general use of "artificially" in this connection is somewhat arbitrary, since it must be assumed that precisely the same transformations occur under "natural" conditions in the stars.

[b] Usually, to place generalisations before data is a procedure not to be recommended in a book on experimental physics. In the present instance, however, the data are already so extensive that to consider them without first establishing certain guiding principles is clearly unprofitable; on the other hand it must continually be borne in mind that the very terms in which these principles are enunciated themselves presuppose considerable familiarity with the results of experiment. The "expectations" here given were not formulated at any very early stage in the history of the subject.

TYPES OF ARTIFICIAL TRANSFORMATION

We have already (p. 42) made the distinction between transformations which involve capture of the incident projectile and those which do not. Let us consider, first, transformations of the latter type. They may result in the emission either of radiation only, or of particles (with or without additional radiation). If radiation only is emitted it may be characterised by a continuous (frequency) spectrum or by a line spectrum. Strictly we should regard a non-capture process resulting merely in radiation as a nuclear transformation only when the radiation is of a definite frequency (or of definite frequencies). Then it may be assumed to arise in the recovery of the nucleus from a state of excitation, otherwise (when the spectrum of radiation is continuous) it must be taken to correspond to the white X-radiation in the analogous electron case. When particles and radiation are observed together, again radiation of definite frequencies should generally be interpreted in terms of the excitation of the nucleus which remains after the emission of the particle. Conversely, when particles are observed, but no radiation, clearly the product nucleus is always left in the ground state.

The same general possibilities likewise characterise those transformations which involve capture of the incident particle: this particle may simply be captured, with the emission of radiation, or a particle or particles may be ejected from the unstable system thus synthesised. In the latter case excitation may or may not result, for any of the complex products of disintegration: the emission of radiation (having a discrete energy spectrum)[a] pointing to the former, its absence to the latter alternative. Furthermore, when such excitation is in question, we might expect, on general grounds, that affirmative evidence should be provided as much by the energies of the particles as by the occurrence of quanta. Thus our ideas of conservation lead to the conclusion that the product nucleus, for example, may remain excited only when a particle of less than the normal energy has been emitted. Already a brief formal discussion of this point of view has been given in § 16; its relevance in respect of spontaneous

[a] See, however, pp. 174, *et seq.*

radioactive disintegrations has been fully demonstrated in § 31.

Investigations of artificial disintegration, therefore, should direct attention to the following points: the nature of the corpuscular products, the balance of kinetic energy in the process and the question of associated quantum radiations. They should also be concerned with another question which, though not material to the classification of the nuclear change, is very important for any consideration of its mechanism: they should include a determination of the disintegration yield as a function of the energy of the responsible particle. The importance of such a determination has already been noticed in an earlier discussion (p. 51); it has to do, more particularly, with the phenomenon of "resonance disintegration".

Finally, before leaving generalities entirely, we may notice a shortened type of formal transformation scheme, cf. (14), p. 81, which may be useful in certain connections. Bothe[a] has suggested the notation

$$ {}_{e}X^{a}\ ({}_{f}x^{b},\ {}_{h}y^{d})\ {}_{g}Y^{c} \qquad \ldots\ldots(31)$$

to replace the longer form, more generally employed. There can be no doubt but that (31) represents satisfactorily the standard capture disintegration in which two initial and two final particles are concerned—as long as the energetics of the change are not in question.

§ 33. *Disintegration with the emission of protons.* The transformations to be discussed in this section are in many respects more completely investigated than those of any other type. In spite of various suggestions to the contrary,[b] with perhaps one exception,[c] they are simple capture disintegrations, all of which may be regarded as special cases of the general transformation

$$ {}_{Z}X^{A}\ ({}_{2}\mathrm{He}^{4},\ {}_{1}\mathrm{H}^{1})\ {}_{Z+1}Y^{A+3} \qquad \ldots\ldots(32).$$

[a] Bothe, *Phys. Z.* 36, 776, 1935.

[b] Cf. Chadwick and Gamow, *Nature*, 126, 54, 1930.

[c] The disintegration, ${}_{1}\mathrm{H}^{2}\ (+{}_{2}\mathrm{He}^{4}) \rightarrow {}_{1}\mathrm{H}^{1} + {}_{0}n' + ({}_{2}\mathrm{He}^{4})$ (Dunning, *Phys. Rev.* 45, 586, 1934).

In the earlier experiments[a] the elements disintegrated appeared to be always of odd atomic number ($Z=2n+1$), and (with the exception of nitrogen) of mass number (A) of the form $4n+3$.[b] Further investigation, however, has shown that no real limitation is actually involved; the three isotopes of magnesium, for example, all of which are susceptible of disintegration in this way, are of even atomic number and represent the mass types $4n$, $4n+1$, $4n+2$, respectively. Moreover, there are many other similar cases. Yet the apparent regularity was not entirely illusory. It will readily be understood that disintegration is most easily detected when the release of energy is considerable; now it appears that this release is generally large when the mass number of the original nucleus is of the type $4n+3$—nuclei of this class are characterised by relatively small energies of binding.

We may conclude then that very many light nuclei, of varied mass types, are susceptible of disintegration—and, because the data which refer to them are collected together in § 36, we may choose a particular example, only, for further discussion. Actually it is not easy to find a single element which is completely satisfactory for this purpose: we shall, therefore, discuss in turn the (proton) disintegrations of boron and aluminium. Recently the protons emitted in the former transformation have been fully studied by Miller, Duncanson and May[c] and by Paton,[d] and those produced in the latter process by Duncanson and Miller[e] and by Haxel,[f] amongst others.

Fig. 15 represents the results of Paton with boron—reinterpreted slightly after comparison with the data of Miller, Duncanson and May. The curve shows the decrease, with increasing thickness of absorbing matter, in the number of protons observed within a narrow cone of directions at about

[a] Rutherford and Chadwick, *Phil. Mag.* 42, 809, 1921.

[b] The results with boron are now known to have been misinterpreted in this particular.

[c] Miller, Duncanson and May, *Proc. Camb. Phil. Soc.* 30, 549, 1934.

[d] Paton, *Z. Physik*, 90, 586, 1934; *Phys. Rev.* 46, 229, 1934.

[e] Duncanson and Miller, *Proc. Roy. Soc.* 146, 396, 1934.

[f] Haxel, *Z. Physik*, 88, 346; *ibid.* 90, 373, 1934.

90° to the axis of a beam of the full range α particles from thorium C′, the target being of small air equivalent thickness. It will be observed that the arrangement is one which is particularly suited to the accurate determination of the balance of energy: the α particles are initially all of the same velocity[a]—and in general lose very little energy in passing through the target—and the uncertainty in θ, the angle between the directions of motion of captured α particle and

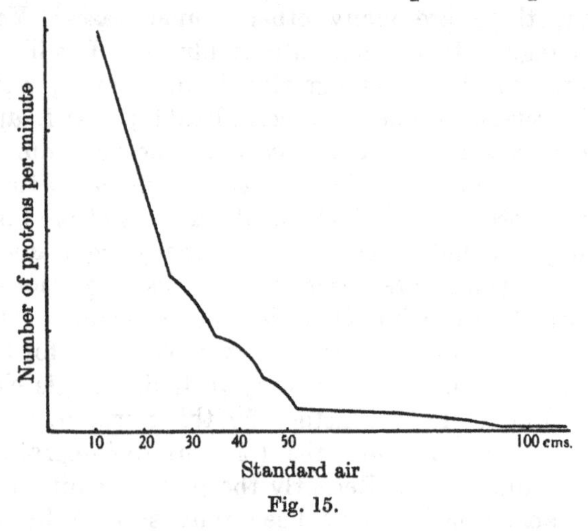

Fig. 15.

ejected proton, is relatively small. Equation (18), p. 82, may therefore be applied. It is evident from the figure that at least five different values of Q, the amount of kinetic energy released, are in question; the only ambiguity concerns the mass of the nucleus disintegrated. Boron is a mixed element of mass numbers 10 and 11; disintegration would lead to the nuclear products C^{13} and C^{14}, respectively. We may say at once that the proton group of longest range must definitely[b] be ascribed to the former alternative, ${}_5B^{10}$ (${}_2He^4$, ${}_1H^1$) ${}_6C^{13}$.[c] A similar conclusion regarding the group next in order of

[a] Complication due to the α particles of thorium C, also present in this case, was shown, in fact, not to arise.

[b] Chadwick, Constable and Pollard, *Proc. Roy. Soc.* 130, 463, 1931.

[c] If this were not so the mass of C^{14} would be so much smaller than that of N^{14} that the latter element would be unstable in respect of positron emission.

range also rests on sound evidence (*v. inf.*) and altogether at present[a] there is no reason to conclude that any of the observations require the assumption of the latter transformation, ${}_5B^{11}$ (${}_2He^4$, ${}_1H^1$) ${}_6C^{14}$, for their complete explanation. Retaining the former interpretation throughout, the appropriate values of Q are $+3{\cdot}1$, $+0{\cdot}4$, $-0{\cdot}1$, $-1{\cdot}0$ and $-1{\cdot}9\times10^6$ electron volts, respectively. Even if we cannot be certain that the greatest of these corresponds to the formation of the residual nucleus in the ground state,[b] there are still many possibilities for the emission of quanta. Precisely because the short range proton groups are generally the more intense, we should expect the emission of γ radiation to be a fairly frequent occurrence also. The most important observations in this connection have been made by Becker and Bothe[c] and by v. Baeyer.[d] The former workers employed coincidence counters for the investigation of the recoil electrons produced by this γ radiation and by comparative absorption measurements deduced an effective quantum energy of $3{\cdot}1\times10^6$ electron volts for it;[e] the latter investigator, observing the coincident discharges of counters arranged to record protons and recoil electrons, respectively, showed that the correlation between the emission of short-range protons and the emission of quanta was much closer than that relating the emission of protons of long range with any quantum emission. It will be evident that these two investigations, together, place upon a very sound basis the general interpretation of the several groups of protons observed, within a small solid angle, when

[a] Since C^{14} would be expected to be radioactive, so long as this activity is not found the primary disintegration will continue to be assigned to B^{10}. It may be noted, however, that the masses given in Table 6 do not exclude the possibility of the disintegration of B^{11}, even if C^{14} is radioactive, with a positive value of Q. [C^{14}, obtained by bombarding carbon with deuterons, has recently been shown to be radioactive (McMillan, *Phys. Rev.* 49, 875, 1936).]

[b] The masses of Table 6 make it likely that transition to the ground state involves a release of energy of $4{\cdot}3\times10^6$ electron volts. This transformation is clearly very improbable according to Fig. 15. Here the residual counts beyond 100 cm. absorption are certainly due in part to the effects of neutrons (§ 34).

[c] Becker and Bothe, *Z. Physik*, 76, 421, 1932.

[d] v. Baeyer, *Z. Physik*, 95, 417, 1935.

[e] This is the sound evidence previously (p. 156) spoken of as favouring a common origin for the proton groups for which Q values are $+3{\cdot}1$ and $+0{\cdot}4$ (or $-0{\cdot}1$) $\times10^6$ electron volts, respectively.

monoenergetic α particles traverse a thin layer of material which is capable of disintegration. It was for this very reason that the case of boron was chosen as the first example; evidence concerning the emission of γ radiation is here most precise. On the other hand there are two reasons why the disintegration of boron is not a good choice when consideration of the variation of disintegration yield with primary energy is in question. In the first place evidence from scattering experiments[a] shows that α particles of about $3{\cdot}6 \times 10^6$ electron volts energy and above pass without great hindrance into the nuclear structure, thus no important variations of disintegration efficiency are to be expected for α particles of energy greater than this, and, secondly, it is not easy to obtain beams of α particles of much lower energy free from the "straggling" which makes much less definite any resonance phenomenon. Nevertheless, as indicated in the collected results (p. 168), such phenomena have been found, even with boron, by Pollard[b] and by Miller, Duncanson and May. With aluminium they are altogether more in evidence; for that reason we shall proceed at once to a discussion of the experimental results in this case. The work of Duncanson and Miller and of Haxel, to which preliminary reference has already been made, may here be regarded as completing the pioneer investigations of Pose[c] and of Chadwick and Constable:[d] by it these investigations were extended in that the energy of the α particles employed was greater than had previously been possible. Fig. 16 attempts to combine all the results in a single curve. It may be thought of as representing the yield of protons, say of the two groups of greatest energy,[e] as a function of the energy of the α particles incident upon a thin layer of aluminium. It shows the existence of six favoured

[a] Riezler, *Proc. Roy. Soc.* 134, 154, 1931.

[b] Pollard, *Phys. Rev.* 45, 555, 1934.

[c] See Diebner and Pose, *Z. Physik*, 75, 753, 1932.

[d] Chadwick and Constable, *Proc. Roy. Soc.* 135, 48, 1932.

[e] A little consideration will show that spurious effects may be obtained if proton groups be included for which Q is much less than zero. Then, within the range covered by the experiments, protons of these groups will cease to be recorded—from energy insufficiency alone—when the range of the bombarding α particles is suitably reduced (cf. p. 82).

energies for which resonance disintegration is possible and it indicates the rapid increase of disintegration probability as the energy of the α particle becomes comparable with the energy summit of the potential barrier (p. 51). Above this energy (about $7{\cdot}5 \times 10^6$ electron volts), however, it appears that the probability of disintegration does not further increase.

Let us consider now another possibility in view of the occurrence of resonance phenomena such as have been described. Suppose that a thick layer of material is bombarded with α particles of initial energy which does not allow them

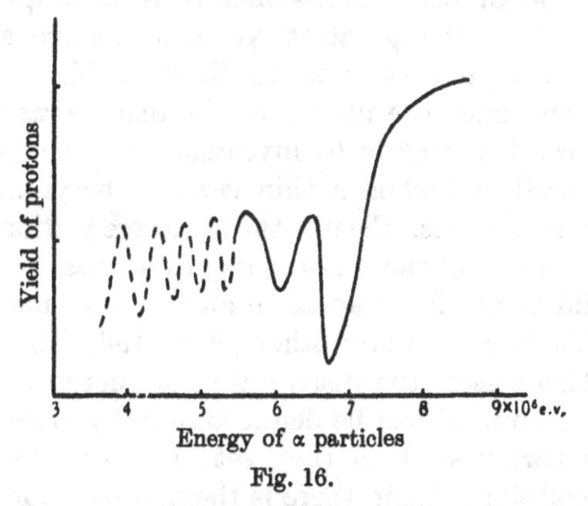

Fig. 16.

to penetrate directly into the nuclear structure (as, for example, the α particles from polonium are unable to enter the aluminium nucleus, except in conditions of resonance). Then, for a single value of Q, it is still possible that distinct groups of protons may be observed (in a given direction). In actual fact, it was in experiments of this kind that evidence for α particle resonance was first obtained (cf. p. 43). Clearly, if several values of the balance of kinetic energy are possible (say, m such values), then the number of proton groups may reach a total of mn, if resonance occurs for n separate energies of the bombarding particle.

The data of § 36 refer to the resonance energies, potential summit energies and the energies of reaction (Q values) in the known cases of proton disintegration produced by α particle bombardment. They should not now require further explanation.

§ 34. *Disintegration with the emission of neutrons.* In this section we shall again consider, first of all, a particular case in some detail and then proceed to discuss, in more general terms, the fact that, with many elements, bombardment by α particles results in the emission of both protons and neutrons—representing alternative modes of disintegration (so it may be shown) of nuclei of a single type.

As concerns the particular case, the obvious choice is the disintegration of beryllium. Effectively a simple element, beryllium gives the greatest yield of neutrons—and no protons—for α particles of ordinarily available energies. The first problem, once the nature of the disintegration process has been established,[a] is to investigate the energies of the neutrons produced when a thin layer of beryllium is bombarded by α particles. Clearly the same precautions concerning the definition of the initial energies of these particles and of the solid angles for irradiation and observation are necessary in this case as in any other (cf. p. 156). Then a further complication arises: the distribution of energy amongst the neutrons so emitted can be deduced only by observing other (charged) particles which they set in motion by elastic or inelastic collision. Again there is the necessity for the definition of angular conditions in respect of these encounters, with the corresponding loss in intensity or resolution, or both. Inelastic (disintegration) collisions will be discussed in the next chapter, here we shall be concerned entirely with the evidence from elastic collisions. On account of the near equality in mass between neutron and proton, a proton projected within a small angle of the direction of motion of a neutron which collides with it takes almost the whole of the energy of the neutron. The ideal method of experiment, therefore, consists in obtaining a range distribution curve for the protons projected "forwards" from a thin layer of paraffin wax by the neutrons emitted within a small solid angle from a

[a] Chadwick, *Proc. Roy. Soc.* 136, 692, 1932.

thin beryllium target irradiated by a roughly canalised beam of mono-energetic α particles. Another method, of some convenience, is to irradiate the gas in an expansion chamber with the neutrons and measure the ranges (and the angles of projection) of the (relatively heavy) nuclei set in motion in this way.[a] If a high pressure expansion chamber be used,[b] hydrogen gas or methane may be employed and the proton tracks measured, to great advantage. Adopting the paraffin wax—ionisation chamber arrangement, with polonium α particles and a thin beryllium target, Chadwick[c] obtained results which are given in Fig. 17. The two curves represent

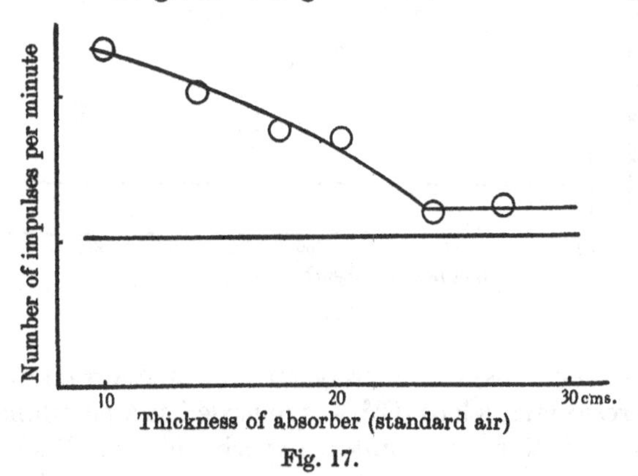

Fig. 17.

the numbers of particles traversing the chamber, per unit time, with and without the paraffin in position. It is clear that a main group of protons of range about 24 cm. in air, and a weaker group (or groups) of higher energy, are indicated. On account of intensity considerations a satisfactory investigation of these high energy components is extremely difficult. Probably the most accurate results are those of Dunning[d] who used the complex α radiation from very intense radon sources ($\sim$ 1000 millicuries) to bombard

[a] Feather, *Proc. Roy. Soc.* 142, 689, 1933; Monod-Herzen, *Ann. Physique*, 4, 137, 1935.

[b] Cf. Bonner and Mott-Smith, *Phys. Rev.* 46, 258, 1934; Bonner and Brubaker, *ibid.* 48, 742, 1935.

[c] Chadwick, *Proc. Roy. Soc.* 142, 1, 1933.

[d] Dunning, *Phys. Rev.* 45, 586, 1934.

powdered beryllium and, after some 200 hours recording, obtained the data given in Fig. 18. Considering Figs. 17 and 18 together we may say that energy changes (Q) of $6{\cdot}6 \pm 0{\cdot}5$ and $-0{\cdot}5 \pm 0{\cdot}2 \times 10^6$ electron volts are involved. Evidence concerning the emission of γ radiation is in reasonable agreement with this, particularly if an intermediate group of

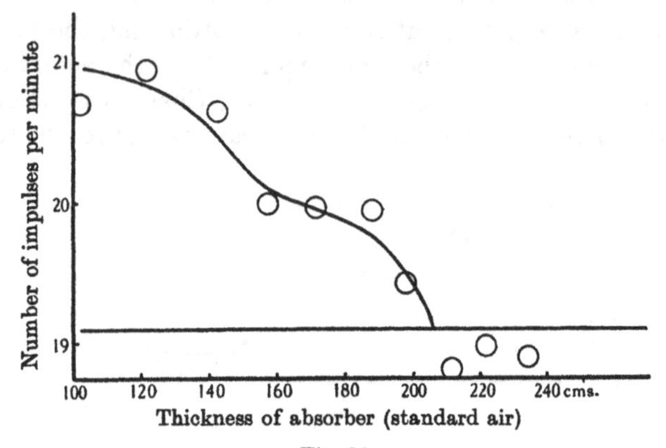

Fig. 18.

protons (neutrons)—or merely an intermediate energy level for the residual nucleus (C^{12})—is regarded as a possibility. The original coincidence counter experiments of Becker and Bothe[a] suggested an effective quantum energy of $5{\cdot}1 \times 10^6$ electron volts and, if very similar data have more recently been differently interpreted,[b] a small number of high energy electron tracks found in the expansion chamber[c] can only be attributed to γ rays of about 7×10^6 electron volts energy.[d]

[a] Becker and Bothe, *Z. Physik*, 76, 421, 1932.

[b] Bernardini and Emo, *La Ricerca Scientifica*, 6 (2), 17, 1935.

[c] Auger, *Comptes rendus*, 194, 877, 1932; Chadwick, Blackett and Occhialini, *Proc. Roy. Soc.* 144, 235, 1934. See, also, Koch and Rieder, *Sitzungsber. Akad. Wiss. Wien*, II*a*, 144, 331, 1935.

[d] For other evidence concerning the excited states of the nucleus C^{12} see Crane, Delsasso, Fowler and Lauritsen, *Phys. Rev.* 48, 100, 1935; Cockcroft and Lewis, *Proc. Roy. Soc.* 154, 261, 1936. (More accurate experiments (Bothe, *Z. Physik*, 100, 273, 1936) have now shown that the γ radiation from beryllium bombarded by α particles has three components, of $2{\cdot}7$, $4{\cdot}2$ and $6{\cdot}7 \times 10^6$ electron volts energy. This represents very good agreement with the data concerning neutron energies. For further details see Maier-Leibnitz, *Z. Physik*, 101, 478, 1936.)

One further point should be noticed before we proceed; the use of the range distribution curve for the protons from paraffin in this way to make predictions concerning the γ radiation to be expected from the primary neutron disintegration produced by α particles is complicated by a fact which did not enter into previous considerations of a similar nature (p. 157). Ideally, for the proton disintegrations discussed in § 33, the range distribution curve provides information concerning both energies and intensities of the quantum radiations; in the neutron case, however, because of the variation of the neutron-proton collision cross-section with velocity (favouring the detection of low energy neutrons), only the most tentative conclusions concerning intensities may be made. At present this is not a serious difficulty, other data regarding intensities being equally meagre,[a] but in principle it represents an important distinction.

At this stage, following our accepted scheme of discussion (p. 154), we should refer briefly to the investigations which have been made of the variation of disintegration yield with α particle energy. Here the results of a large number of experimenters[b] are in substantial agreement. The potential barrier of the beryllium nucleus for α particles appears to be easily penetrable for energies greater than $3{\cdot}5 \times 10^6$ electron volts and again, by a resonance effect, for energies neighbouring upon 2·5 and $1{\cdot}6 \times 10^6$ electron volts. Furthermore, the intensity of the γ radiation has been found to exhibit variations which may be similarly described.[c] Fig. 19, uncorrected for the variations in neutron-proton collision cross-section referred to above, may be taken as characteristic of the general results.

Having discussed the two types of disintegration separately we may finally sum up the situation by saying that, whether neutrons or protons are emitted as the result of α particle bombardment, essentially the same general phenomena occur;

[a] Cf. Lea, *Proc. Roy. Soc.* 150, 637, 1935.

[b] E.g. Rasetti, *Z. Physik*, 78, 165, 1932; Curie and Joliot, *J. Physique*, 4, 278, 1933; Chadwick, *Proc. Roy. Soc.* 142, 1, 1933; Bernardini, *Z. Physik*, 85, 555, 1933; Emo, *Nuovo Cimento*, 11, 357, 1934.

[c] Becker and Bothe, *loc. cit.*

the energy spectra of disintegration particles and quanta and the details of penetration of the nuclear structure by the α particle which produces the transformation run strictly parallel in the two cases. As already mentioned, frequently a single species is susceptible of disintegration in either way; then the comparison may be more closely drawn. This is our next consideration.

It became clear, at an early stage, that the simple elements fluorine and aluminium[a] were amongst those from which

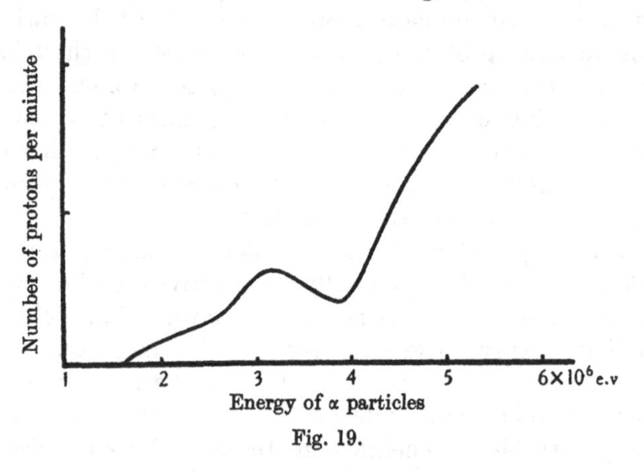

Fig. 19.

neutrons could be obtained under α particle bombardment. Later, sodium provided the third example of a nuclear species for which alternative modes of disintegration had necessarily to be assumed (these three elements, as we have already seen, were known to emit protons under α particle bombardment). In general terms the alternative modes of disintegration may be written as follows:

$$(a)\ {}_{Z}X^{A}\ ({}_{2}\mathrm{He}^{4},\ {}_{1}\mathrm{H}^{1})\ {}_{Z+1}Y^{A+3},\quad (b)\ {}_{Z}X^{A}\ ({}_{2}\mathrm{He}^{4},\ {}_{0}n^{1})\ {}_{Z+2}Y^{A+3} \qquad \ldots\ldots(33).$$

At first[b] these two modes were discussed on the supposition that both product species, ${}_{Z+1}Y^{A+3}$ and ${}_{Z+2}Y^{A+3}$, were stable (as against β disintegration). Immediately this involved certain

[a] Curie and Joliot, *Comptes rendus*, 196, 397, 1933.
[b] Chadwick, *Proc. Roy. Soc.* 142, 1, 1933.

difficulties[a] (no example of two stable isobaric nuclei with consecutive atomic numbers was at that time known amongst the lighter elements),[b] however a natural explanation was provided by the discovery of artificial radioactivity (§ 26). With fluorine, sodium and aluminium the products of proton disintegration, Ne^{22}, Mg^{26} and Si^{30}, were known to be stable; it appears that the corresponding neutron products Na^{22}, Al^{26} and P^{30} are positron emitters. Many other similar cases, not so immediately obvious, have now been investigated. Sometimes the unstable product is negative electron active; then it is the proton product, and the neutron product is stable. As will be observed later (§ 41), a similar classification is possible when deuterons replace α particles as the particles responsible for producing the disintegrations.[c] Another classification depends upon the nuclear type of the element bombarded. In the α particle case, when this species is such that $A=2Z$, invariably the neutron product is unstable (mass number being less than twice the atomic number for this product) and thus the artificial radioactivity is of the positron type, but, when $A=2Z+1$ initially, different results may be obtained, depending upon whether Z is odd or even. When Z is odd, again the neutron product is generally unstable; when Z is even, however, the proton product is radioactive and emits negative electrons. Similar conclusions apply when $A=2Z+2$, but so far only one species of this type ($_{12}Mg^{26}$) has been successfully investigated.

Passing over generalities, then, we may notice one interesting practical result: if need be we may confine attention to the radioactivity of the unstable product, rather than to the heavy particles previously emitted, if our object is to investigate certain features of the primary disintegration.[d] In

[a] Difficulties, of an equally fundamental type, are now known to attach to the alternative supposition then current that a product species might be unstable in the sense that it should transform into its isobar by capturing an electron from the extranuclear system (cf. p. 188).

[b] There seems now no good reason why the disintegrations $_3Li^7$ ($_2He^4$, $_1H^1$) $_4Be^{10}$ and $_3Li^7$ ($_2He^4$, $_0n^1$) $_5B^{10}$ should not both be observed, but so far only the latter has been reported.

[c] This later discussion should also be referred to for a consideration of the balance of energy in the alternative modes of disintegration.

[d] This may be a considerable advantage when the heavy particles are neutrons.

particular we may investigate the dependence of disintegration yield on α particle energy in this way. Amongst others, Ellis and Henderson,[a] Haxel[b] and Fahlenbrach[c] have made such experiments. Comparing the results of these experiments with those concerning the production of the corresponding stable species in the alternative mode, in general it may be said that the same resonance levels are indicated whichever primary process is in question. This is natural enough; these two processes are to be regarded as alternative ways in which the unstable aggregate of original nucleus and captured α particle may possibly break up. Capture of the α particle (of which resonance phenomena may be a feature) is necessary for either process. If, however, the same resonance levels are indicated in the two cases, the branching ratio may (and does)[d] vary with the energy of the α particles. The more endothermic of the primary disintegrations will probably be favoured the higher the energy of the α particles employed. With nitrogen this effect is particularly noticeable: no neutrons—and no artificially radioactive product—are produced when the α particles have less than $5{\cdot}9 \times 10^6$ electron volts energy. For more energetic α particles, on the other hand, this process increases in probability at the expense of the well-known proton disintegration.

§ 35. *Non-capture excitation.* Experimentally the problem of investigating the possible occurrence of non-capture excitation is a difficult one. Since almost all elements undergo capture disintegration to some extent, and the production of excited product nuclei (and thus, subsequently, of γ radiation) is a common feature of such transformations, the problem becomes generally that of discriminating between the γ radiation produced in this way and that due to possible non-capture excitation of the original nucleus. If the characteristic quantum energies were known beforehand it might not be so difficult; when they are not known, or when a number of

[a] Ellis and Henderson, *Proc. Roy. Soc.* 146, 206, 1934.
[b] Haxel, *Z. Physik*, 93, 400, 1935.
[c] Fahlenbrach, *Z. Physik*, 94, 607, 1935.
[d] Haxel, Fahlenbrach, *loc. cit.*

possible energies is predicted, one must look to some special circumstance if simple excitation is to be established. The most clear-cut evidence is obtained when γ radiation can be detected even though the energy of the incident α particles is already so small that capture disintegration (with excitation) is impossible. Such evidence has been obtained with targets of lithium,[a] nitrogen,[b] fluorine and aluminium,[c] respectively. It may be that simple excitation is, in fact, a common process, lacking recognition only because of the experimental difficulties which have been mentioned. Recently, Schnetzler[d] has made a thorough study of simple excitation in the case of lithium. Ascribing both effects to the more abundant isotope, Li^7, it appears that whilst γ radiation is observed first for α particles of roughly 2×10^6 electron volts energy, neutrons are not produced until the energy of the α particles has been increased to 5×10^6 electron volts.[e]

§ 36. *Collected results.*

(*a*) Non-capture excitation:

Table 14

Nucleus	Li^7	N^{14}	F^{19}	Al^{27}
γ ray energies: e.v. $\times 10^{-6}$	0·6, 0·4, (0·2)	?	(1·1), 0·6	?

(*b*) Non-capture disintegration:

$$_1H^2\ (+{}_2He^4) \rightarrow {}_1H^1 + {}_0n^1\ (+{}_2He^4);$$

Energy threshold, $\sim 6{\cdot}5 \times 10^6$ e.v.

(*c*) Radiative capture:

Not known at present.

(*d*) Capture disintegration.

(i) Conditions of entry of α particle into target nucleus.

[a] Bothe and Becker, *Z. Physik*, 66, 289, 1930; Webster (H. C.), *Proc. Roy. Soc.* 136, 428, 1932.

[b] Kara-Michailova, *Sitzungsber. Akad. Wiss. Wien*, II*a*, 143, 1, 1934.

[c] Savel, *Ann. Physique*, 4, 88, 1935.

[d] Schnetzler, *Z. Physik*, 95, 302, 1935.

[e] The problem of "white" γ radiation produced by inelastic (non-excitation) collisions between α particles and nuclei has recently been investigated by Lewis and Bowden (*Phil. Mag.* 20, 294, 1935) with negative results.

Table 15

Nucleus	E_0 e.v. × 10^{-6}	E_r e.v. × 10^{-6}
Li^7	3·0	—
Be^9	3·5	2·5, 1·6
B^{10}	3·6	3·0
B^{11}	3·7	2·4
N^{14}	4·1	3·5
F^{19}	5·0	4·0, 3·4
Na^{23}	—	At least one value
Mg^{24}	6·5	6·3, 5·7
Al^{27}	6·8	6·6, 5·75, 5·25, 4·85, 4·5, 4·0

E_0: Minimum energy for penetration of α particle into nuclear structure with reasonable probability. (E_0 is thus somewhat less than the energy corresponding to the potential summit.)

E_r: Energies for resonance penetration.

(ii) Energy balance in various transformations.

Disintegrations of the type ${}_ZX^A$ (${}_2He^4$, ${}_1H^1$) ${}_{Z+1}Y^{A+3}$.

Table 16

Target nucleus	Q e.v. × 10^{-6}	Product nucleus
${}_5B^{10}$	3·1, 0·4, −0·1, −1·0, −1·9	${}_6C^{13}$
${}_7N^{14}$	−1·3	${}_8O^{17}$
${}_9F^{19}$	1·5, −0·1, −2·1, −3·3	${}_{10}Ne^{22}$
${}_{11}Na^{23}$	1·9, −0·4, −2·1, −3·1	${}_{12}Mg^{26}$
${}_{12}Mg^{24}$	−1·2, −2·0, −2·9	${}_{13}Al^{27}$
${}_{13}Al^{27}$	2·1, −0·2, −1·5, −2·8	${}_{14}Si^{30}$
${}_{14}Si^{28}$	−2·4, −3·2, −4·0	${}_{15}P^{31}$
${}_{15}P^{31}$	(?), −0·1, −1·4, −3·0	${}_{16}S^{34}$
${}_{16}S^{32}$	−2·3, −2·9, −3·7	${}_{17}Cl^{35}$
${}_{17}Cl^{35}$	0·1, −2·4, −4·0	${}_{18}A^{38}$
${}_{19}K^{39}$	(?), −0·9, −2·3, −3·4	${}_{20}Ca^{42}$

Also, (Q values not known) Ne^{20}, Mg^{25}, Mg^{26}, Si^{29}, A^{40} and Ca^{40} undergo disintegration of this type.

Disintegrations of the type ${}_ZX^A$ (${}_2He^4$, ${}_1H^2$) ${}_{Z+1}Y^{A+2}$.

The possibility of the occurrence of disintegrations of this type has been discussed by a number of authors[a] and experi-

[a] Perrin, *Comptes rendus*, 194, 2211, 1932; Cockcroft and Lewis, *Proc. Roy. Soc.* 154, 261, 1936.

ments have been carried out particularly to search for them, but so far without success.[a]

Disintegrations of the type $_{Z}X^{A}$ ($_{2}He^{4}$, $_{0}n^{1}$) $_{Z+2}Y^{A+3}$.

Table 17

Target nucleus	Q e.v. $\times 10^{-6}$	Product nucleus
$_{3}Li^{7}$	−3·0	$_{5}B^{10}$
$_{4}Be^{9}$	6·2, −0·5	$_{6}C^{12}$
$_{5}B^{11}$	−0·6 and other values	$_{7}N^{14}$
$_{7}N^{14}$	−4·6	$_{9}F^{17}$
$_{9}F^{19}$	−2·4 and other values	$_{11}Na^{22}$

Also, (Q values not known) B^{10}, Ne^{22}, Na^{23}, Mg^{24}, Al^{27}, P^{31} and K^{41} undergo disintegration of this type.

Q: Kinetic energy released in the transformation. According to the accepted interpretation, energies of excitation of the product nuclei may be obtained by subtracting from the largest value of Q, in each case, each of the smaller Q values. Thus $0{\cdot}8 \times 10^{6}$ electron volts is a possible excitation energy for the nucleus Al^{27} (Table 16). It seems probable that the aluminium nucleus may be raised to this state of excitation by inelastic (non-capture) collision (Table 14). Again, referring to Table 16, it will be observed that product nuclei of the mass types $4n+1$, $4n+2$ and $4n+3$ are represented. It appears that to each mass type belongs a characteristic set of excited states,[b] the difference of energy between the first excited state and the ground state being greatest when $A=4n+1$[c] and least when $A=4n+3$. This is in general accord with the result that three of the four species for which non-capture excitation has been established fall in the last class, whilst for the fourth such species $A=4n+2$. Non-capture excitation is more easily detected when the energy of excitation is small.

[a] Pollard and Eaton, *Phys. Rev.* 46, 528, 1934.

[b] Cf. Haxel, *Phys. Z.* 36, 804, 1935; May and Vaidyanathan, *Proc. Roy. Soc.* 155, 519, 1936.

[c] Results regarding the formation of O^{17} difficult to reconcile with those used in Tables 15 and 16 have also been obtained from time to time: see Stegmann, *Z. Physik*, 95, 72, 1935; Pose, *ibid.* 95, 84, 1935; Stetter, *ibid.* 100, 652, 1936.

CHAPTER XI

TRANSFORMATIONS PRODUCED BY NEUTRONS

§ 37. *Transformations produced by high energy neutrons.* There are two ways of surveying briefly the large amount of experimental material in a rapidly expanding subject like the present one. We may attempt to distinguish formally between various types of transformation to be expected and treat the data on the basis of this classification, or we may recognise the empirical classification, which the possibility of different methods of experimenting automatically provides, and confine attention in turn to one or other type of experiment. In the last chapter we adopted the former method, here the latter approach seems definitely to be preferred.

Originally evidence concerning disintegrations produced by neutrons was obtained directly through work with the expansion chamber—and with neutron sources consisting of some target (generally beryllium) bombarded by the α particles from a radioactive preparation (most frequently of polonium). We shall note, as we proceed, the way in which additional information has resulted from the use of more intense sources of neutrons with the expansion chamber (sources obtained by employing deuterons, rather than α particles, as primary projectiles)[a] and from entirely different methods of experiment.

With the expansion chamber the first evidence of disintegration was found by the writer[b] when nitrogen was irradiated. Plate II (*a*) is a reproduction of one of these early disintegration photographs. It is worth while discussing it with reference to the statements already made (p. 154) regarding the objects of any investigation of a case of artificial disintegration. In this particular case (of neutron provoked disintegration) the question of the balance of kinetic energy in the

[a] Cf. § 41. [b] Feather, *Proc. Roy. Soc.* 136, 709, 1932.

process is intimately bound up with what is usually the prior decision concerning the nature of the products of transformation. The two questions must now be considered together; the only decision which it is sometimes possible to make in advance being the decision between capture and non-capture for the incident neutron—but even this decision may not always be made. We are limited here, as throughout the whole investigation, by the fact that there is no cloud track marking out the path of the neutron before collision: we merely know the position and size of the neutron source and the point at which disintegration occurred—and we must bear in mind the possibility that, whilst all neutrons certainly originate in the source, some may undergo large angle scattering before they are effective in producing disintegration. Nevertheless, it is safe to conclude both that this scattering is absent and that neutron capture has occurred when the geometrical plane defined by the pair of tracks which diverge from the point of disintegration also passes through the volume occupied by the source.[a] With many paired tracks this condition is found to be fulfilled: capture disintegration obviously takes place. Whether or not non-capture processes also occur is at present undecided; it appears probable[b] that in most cases, at least, the non-fulfilment of the condition of coplanarity is evidence rather for scattering of the effective neutron than for disintegration without capture. Assuming capture, when the coplanarity condition is fulfilled, we have the lengths of the cloud tracks and the angles of projection (carrying rather large uncertainties on account of the finite size of the source)—and we may make a number of hypotheses regarding the identity of the products (supposing, for example, the emission of a proton, a deuteron, or an α particle from the nuclear aggregate). Deducing, in respect of each hypothesis in turn, particle energies from the observed lengths of track (this procedure again being subject to considerable uncertainty

[a] This criterion, of course, is based on the assumption of conservation of momentum in the collision.

[b] Cf. Harkins and Gans, *Phys. Rev.* 46, 397, 1934.

with the heavy products, for lack of sufficiently extensive range-energy data), we may attempt to decide between the various suggestions by a further application of the principle of the conservation of momentum. The resultant momentum of the products must not only lie in a particular plane (coplanarity condition), but it must be directed through the source and in magnitude it must represent a possible value of the momentum of a neutron emitted by the source. In the case of the disintegration of Plate II (*a*) all these conditions were fulfilled on the assumption of the process

$$_{7}N^{14}+{}_{0}n^{1}\rightarrow{}_{5}B^{11}+{}_{2}He^{4} \qquad \ldots\ldots(34)$$

and on no other possible assumption. The balance of kinetic energy was also given at the same time as this decision was made, the energies appropriate to the—successfully identified—products having been used to calculate the momentum (or energy) of the neutron which was captured. As regards the accuracy of the determination it can only be said that the balance of energy deduced in this way may be subject to an uncertainty probably never less than $0{\cdot}5\times10^{6}$ electron volts and often approaching 2×10^{6} electron volts. Clearly a statistical treatment of results is imperative. A first step would be to include those paired tracks previously excluded by the application of the condition of coplanarity. We might interpret them on the basis of (34), also, and so deduce both the initial energy and the initial direction of motion of the captured neutron. As already indicated, this step does not generally lead to inconsistencies. A much better method, however, is to multiply examples of incontestibly capture events by the use of more intense sources of neutrons. The disintegration of nitrogen has been investigated very thoroughly in this way by Kurie[a] and by Bonner and Brubaker,[b] the former worker using the neutrons obtained by bombarding beryllium by deuterons (p. 199), the latter workers neutrons from a similar source, and, in addition, the more energetic particles from lithium bombarded by deuterons (p. 195). These new results exhibit two features quite

[a] Kurie, *Phys. Rev.* 47, 97, 1935.
[b] B nner and Brubaker, *Phys. Rev.* 49, 223, 1936.

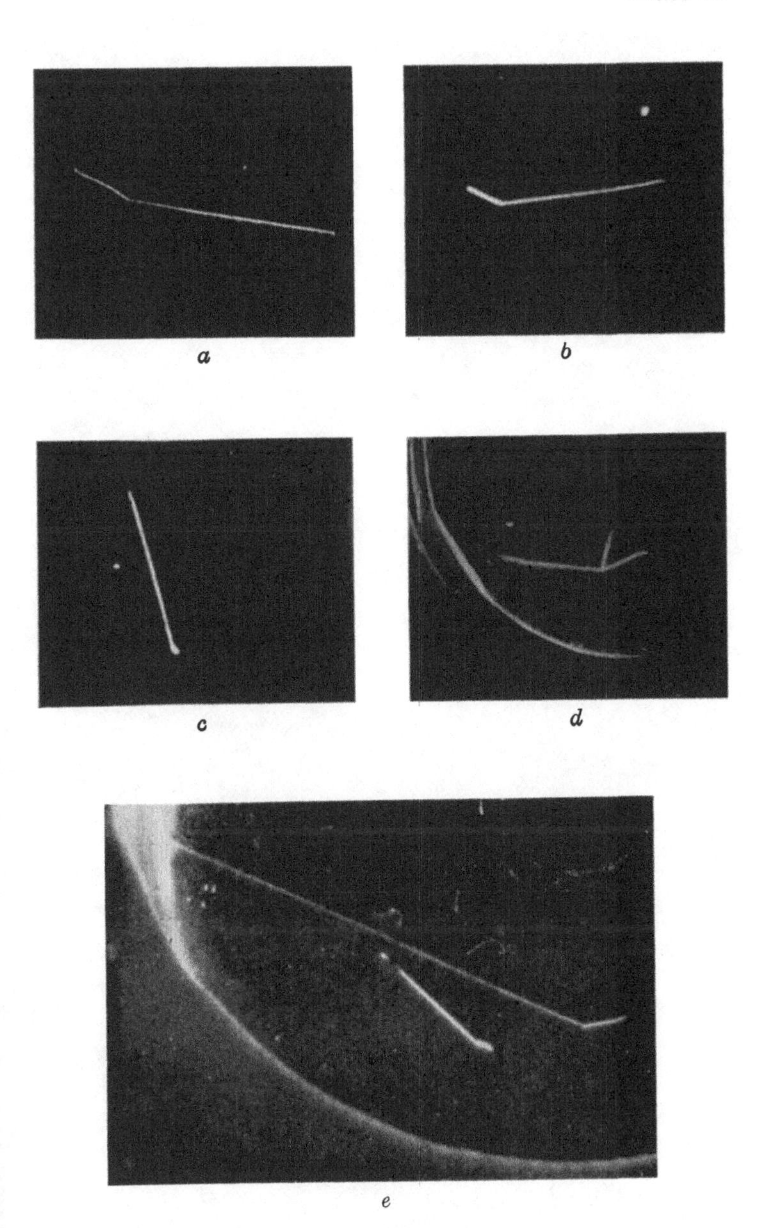

a *b*

c *d*

e

clearly. In the first place, as previously suspected,[a] all the capture disintegrations are not of one type, the disintegration

$$_7N^{14} + _0n^1 \rightarrow _6C^{14} + _1H^1 \qquad ......(35)$$

being represented (see p. 181), in addition to (34),[b] and, secondly, the values of Q for (34) are distributed over a wide range, extending from a possible slight release of kinetic energy (Q positive) to the absorption of 10^7 electron volts energy, at least. We must seek to understand this feature of the results, remembering that even yet the available data are not sufficiently extensive to warrant a clear decision whether the spectrum of Q values is a line spectrum (cf. §§ 33, 34) or whether it is continuous. Obviously our natural bias is towards the former interpretation, but it is as well to be wary of such partisanship. Nevertheless, we may dispose of one aspect of the matter at the outset: the upper limit of Q values appears to agree satisfactorily with what would be deduced on any supposition from the masses of Table 6. The nucleus B^{11} is evidently formed directly in the ground state in a certain fraction of the disintegrations and no γ radiation is emitted. As we shall see (p. 195), the same nucleus is formed, also in the ground state, in the transformation

$$_5B^{10}\,(_1H^2,\, _1H^1)\, _5B^{11},$$

and discrete Q values[c] 9·1, 7·0 and $4{\cdot}6 \times 10^6$ electron volts are generally regarded[d] as evidence for excited states of the nucleus having 2·1 and $4{\cdot}5 \times 10^6$ electron volts energy of excitation. We should naturally look for indications of the same modes of excitation in the neutron provoked transformation—and it may plausibly be maintained that there is some evidence in this direction. But two other considerations must not be forgotten. In the first place higher excitation energies must also be assumed to explain the neutron results—which does not, in itself, introduce any insuperable difficulty

[a] Cf. Feather, *Proc. Roy. Soc.* 142, 689, 1933; Kirsch and Rieder, *Sitzungsber. Akad. Wiss. Wien*, II*a*, 144, 383, 1935.

[b] Examples of paired tracks interpreted in terms of (34) and (35), respectively, are given in (*b*) and (*c*), Plate II. These photographs are from unpublished work by Chadwick, Feather and Davies.

[c] Cockcroft and Lewis, *Proc. Roy. Soc.* 154, 246, 1936.

[d] Cf. Crane, Delsasso, Fowler and Lauritsen, *Phys. Rev.* 46, 1109, 1934.

—and, secondly, there is a peculiar correlation, between the energy absorbed in the disintegration and the energy of the neutron responsible for producing it, which is more difficult to understand. Clearly, a large absorption of kinetic energy is impossible if that energy is not possessed by the incident neutron, but it does appear that cases in which most of the kinetic energy of a high energy neutron reappears in the products of disintegration are sufficiently rare to call for comment.[a] Kurie[b] has attempted an explanation of this fact, assuming that disintegration occurs in two stages. First the neutron is captured, thus radiating a large fraction of its energy—or, at least, making some of this energy available for excitation—then, after a finite time, the new nucleus breaks up with the emission of a heavy particle (in this case an α particle). Certain criticisms concerning points of detail in Kurie's suggestions have already been made,[c] but since these suggestions have many features in common with the ideas of Bohr, to be introduced in § 38, we may give a more formal treatment at this stage. We are considering disintegrations of the general type ${}_ZX^A\ ({}_0n^1, {}_2\mathrm{He}^4)\ {}_{Z-2}Y^{A-3}$.

Let us suppose that the exact masses of neutron and original nucleus, and of the α particle and the final product nucleus, are m, M and m', M', respectively. Let M'' be the mass of the intermediate product, in its state of excitation[d] (supposed always to be the same),[e] and E the (correspondingly unique)

[a] Plate II (e) (Chadwick, Feather and Davies, unpublished), however, provides an instance of this rare occurrence. In this case calculation shows that the captured neutron had kinetic energy of $10{\cdot}2 \times 10^6$ electron volts before the collision and that all except $0{\cdot}4 \times 10^6$ electron volts of this energy appeared afterwards in the products of disintegration ($B^{11} + He^4$).

[b] Kurie, *loc. cit.*

[c] Cf. Feather, *Reports on Progress in Physics*, 2, 74, 1936; Bonner and Brubaker, *loc. cit.*

[d] With nitrogen the intermediate product is N^{15} and the (unstable) state of excitation, postulated by Kurie, one of $14{\cdot}1 \times 10^6$ electron volts excess energy (cf. equation (36) with $E = 3{\cdot}3 \times 10^{-3}$ mass unit and the masses of Table 6). An excited state of $5{\cdot}3 \times 10^6$ electron volts excess energy, indicated by Cockcroft and Lewis (*Proc. Roy. Soc.* 154, 261, 1936) after an investigation of the transformation ${}_7N^{14}\ ({}_1H^2, {}_1H^1)\ {}_7N^{15}$, should not be confused with this hypothetical state. The lower state is definitely stable in respect of particle disintegration (cf. Table 6).

[e] This is an important point of divergence between the suggestions of Kurie and of Bohr.

value of the energy of disintegration (in mass units) appropriate to the formation of the final nucleus in the ground state. Then

$$M'' = M' + m' + E \qquad ......(36).$$

If (cf. equation (16), p. 81)

$$M + m = M' + m' + Q_0 \qquad ......(37),$$

clearly Q_0 is the maximum energy release which would be expected for the complete transformation, normally regarded as energetically a single event. When, as is now assumed, disintegration takes place in two stages, let it be supposed that capture of a neutron of kinetic energy W requires the emission of a quantum or quanta of energy of total amount E_γ. Applying the law of conservation of momentum we have

$$M + m + W = M'' + \frac{m}{M''} W + E_\gamma \qquad ......(38),$$

or, combining with (36) and (37),

$$E_\gamma = \frac{M}{M''} W + Q_0 - E \qquad ......(39).^{a}$$

It is this result which may be used to distinguish between the one-stage and the two-stage hypothesis as here put forward. On the one-stage hypothesis, if Q_0 is positive, disintegration is possible, energetically, for all values of W; if Q_0 is negative, only for initial energies (W) greater than

$$-\frac{M+m}{M} Q_0 \qquad ......(40).^{b}$$

On the two-stage hypothesis, on the other hand, disintegration is impossible unless W is greater than

$$\frac{M''}{M} (E - Q_0) \qquad ......(41);$$

[a] Here the approximation $m + M = M''$ is taken as sufficiently exact.

[b] Again the approximation $M + m = M' + m'$ is employed. The result may be obtained by writing $a = M$, $b = m$, etc., in (18), p. 82. Then a real value of E_y, corresponding to a given angle θ, is obtained only when

$$E_x [(M' + m')(M' - m) + mm' \cos^2 \theta] > - M'(M' + m') Q_0,$$

and, here, E_x is W. When Q_0 is positive, clearly all values of W are permissible, but when Q_0 is negative an absolute lower limit is fixed by $\theta = 0$ and

$$W [(M' + m')(M' - m) + mm'] > - M'(M' + m') Q_0.$$

This reduces to (40) with the approximation indicated.

and, in these limiting conditions, no energy is lost in the form of γ radiation. When $E=0$, obviously (40) and (41) provide essentially the same limitation for W, otherwise the limit is set higher when the two-stage hypothesis is assumed. Following Kurie in that assumption (also in adopting a single value, $3{\cdot}3 \times 10^{-3}$ mass unit for E) we may conclude that the disintegration of nitrogen, ${}_7\mathrm{N}^{14}\,({}_0n^1, {}_2\mathrm{He}^4)\,{}_5\mathrm{B}^{11}$, should be impossible with neutrons of less than $3{\cdot}5 \times 10^6$ electron volts energy ($Q_0 = -0{\cdot}4 \times 10^{-3}$ mass unit). Admittedly it is not so. This fact may be regarded as evidence in favour of the one-stage hypothesis, or it may merely be taken as an indication that the two-stage hypothesis, as here employed, is in need of modification. Formally it may be modified by postulating a number of values for E, rather than a single value; physically such a modification implies a multiplicity in the excited levels of the intermediate product. This brings us nearer to the ideas of Bohr (cf. p. 184), but to the writer, at least, reasons for abandoning the one-stage hypothesis altogether seem as yet to be premature:[a] moreover, even in the two-stage hypothesis, there would appear to be room for the assumption that disintegration of the intermediate product, when it does take place, may result first of all in the production of the final nucleus in an excited state. Evidently no final decision will be possible until much more information is available; here we can leave the matter, therefore, except to remark that the disintegration

$${}_7\mathrm{N}^{14} + {}_0n^1 \rightarrow {}_3\mathrm{Li}^7 + 2\,{}_2\mathrm{He}^4 \qquad \text{......(42)}$$

has recently been observed. Plate II (*d*) is a reproduction of a photograph which Bonner and Brubaker explain in this way. Now the formation of two α particles and a Li^7 nucleus certainly cannot be thought of as an alternative mode of disintegration of the same intermediate product as has been postulated in the more usual event (34): the energy of excitation is too small by about 8×10^6 electron volts.[b] The whole

[a] A detailed theory based upon precisely this hypothesis has been developed by Bethe (*Phys. Rev.* 47, 747, 1935)—apparently with some success.

[b] The position is no better, on this hypothesis, if disintegration into three particles is thought of as taking place in two stages: cf. p. 191.

Plate III

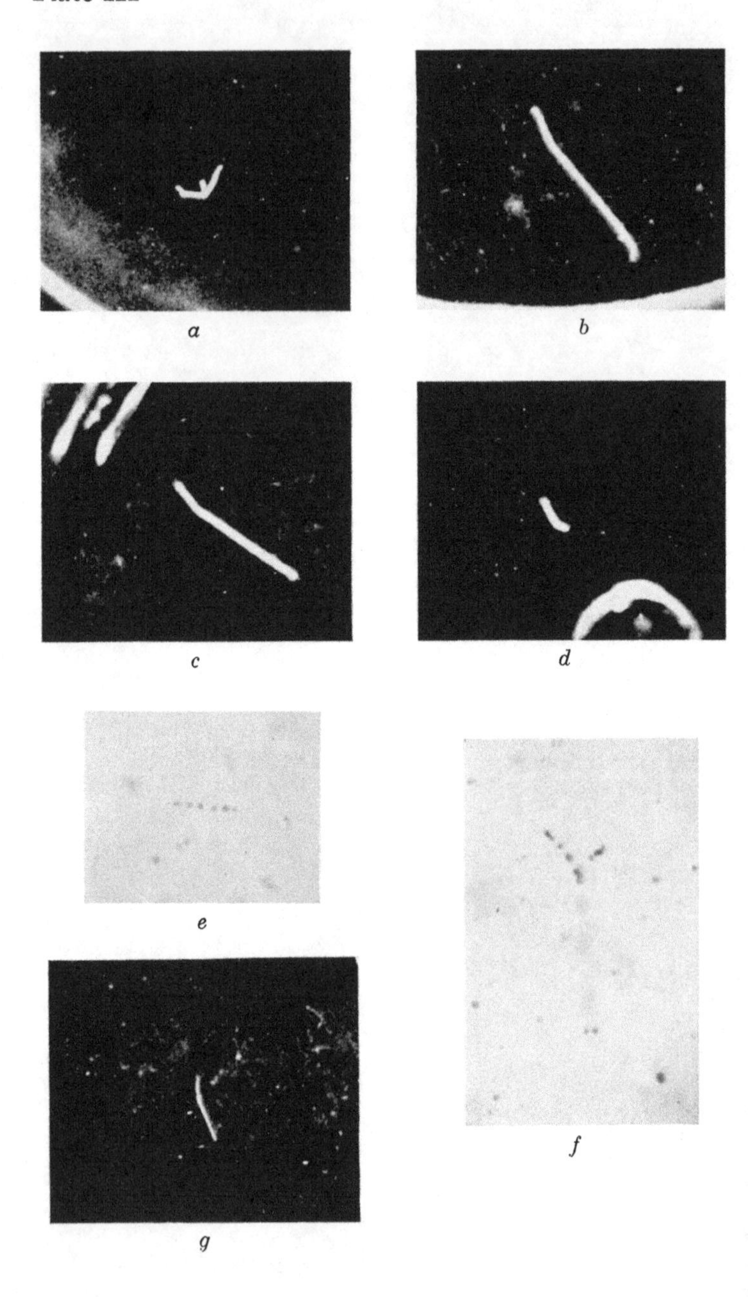

question of alternative modes of transformation (when the corresponding Q_0 values differ widely, one from another) is evidently a serious one for the two-stage hypothesis.

The disintegration of nitrogen having been discussed at this length, other cases must be passed over with a casual mention —and reference to § 39 for the important data. With high energy neutrons carbon, oxygen, fluorine, neon and argon have been investigated and disintegration established with all these elements except the last named. Examples of disintegration photographs are given in Plate III. Plate III (*a*)[a] provides evidence of a three-particle process in the case of carbon: it is probable that the transformation in question is

$$_6C^{12} + {}_0n^1 \rightarrow 2\,{}_2He^4 + {}_2He^5 \qquad ......(43).$$

Normal two-particle disintegrations of oxygen and fluorine are represented by Plate III (*b*) and by Plate III (*c*) and (*d*), respectively.[b] In each case the disintegration particle appears to be an α particle.

Two other methods of experimenting have added to our knowledge of disintegrations produced by neutrons of high energy—the method of the photographic plate (p. 7) in one particular case only, that of artificial radioactivity in a very great number. By exposing a photographic plate impregnated with borax to the action of fast neutrons Taylor[c] obtained evidence for the three-particle process

$$_5B^{10} + {}_0n^1 \rightarrow 2\,{}_2He^4 + {}_1H^3 \qquad ......(44),$$

in addition to the more frequent

$$_5B^{10} + {}_0n^1 \rightarrow {}_3Li^7 + {}_2He^4 \qquad ......(45).$$

Plate III (*f*) is a photo-micrograph of three associated tracks which provide this evidence.[d]

The method of artificial radioactivity springs from the original observation of Fermi (p. 127) that many substances when bombarded by neutrons develop characteristic β

[a] Chadwick, Feather and Davies, *Proc. Camb. Phil. Soc.* 30, 357, 1934.

[b] Chadwick, Feather and Davies, unpublished.

[c] Taylor (H. J.), *Proc. Phys. Soc.* 47, 873, 1935.

[d] In Plate III (*e*) the slow neutron disintegration of B^{10} according to (45) is also represented (see § 38).

particle activities of considerable intensity. Naturally, to investigate these activities with a view to discovering the nature of the primary transformation is a method of wider applicability than that of the expansion chamber, though it is less direct:[a] there is not the restriction to gaseous materials which is there effective. Already the large number of entries in Table 12 (artificially produced electron active species)—most of them representing data obtained under neutron bombardment—is evidence for the very large number of nuclear species which may be transformed in this way. We may say that there is no barrier of potential energy preventing the free entry of the neutron into the structure even of a heavy nucleus.

By chemical methods Fermi and his collaborators have shown that the primary transformation is not always of a single type.[b] Apparently the incident neutron is always[c] captured, but an α particle or a proton[d] may be ejected, or the transformation may be simple capture, only, when the emission of γ radiation must be assumed if conservation of energy is accepted (cf. p. 153). Empirically it is found that in most cases in which heavy elements are activated the last mentioned process is effective. One aspect of this result is not difficult to explain. Whilst it may not be clear why simple capture takes place so efficiently, the decrease in probability of the other (disintegration) processes may easily be understood: as the atomic number of the nucleus increases the chance of a heavy particle escaping, except in conditions of resonance, rapidly becomes vanishingly small (cf. p. 52). It is possible that some of the primary transformations which

[a] Actually, Harkins, Gans and Newson (*Phys. Rev.* 44, 945, 1933), having interpreted the expansion chamber photographs with fluorine (CF_2Cl_2) in terms of the transformation ${}_9F^{19}$ (${}_0n^1$, ${}_2He^4$) ${}_7N^{16}$, suggested that possibly the product nucleus, N^{16}, might be β active, before this result was established, independently, by Fermi.

[b] Fermi, Amaldi, D'Agostino, Rasetti and Segrè, *Proc. Roy. Soc.* 146, 483, 1934.

[c] If exceptions occur, at least they are few in number and at present not clearly established; cf. p. 185.

[d] Chemical methods, of course, cannot be used to distinguish between the emission of a proton and a deuteron. Generally, however, the former is the more probable.

occur with uranium and thorium[a] are to be explained in terms of such a resonance effect in respect of the escaping particle:[b] only further experiment can decide. Here, as already indicated, we are left with the task of explaining the common occurrence of the simple capture transformation. For this process to be reasonably frequent we should normally expect that the emission of radiation must take place in a time comparable with that during which interaction between the nucleus and an incident neutron remains large when the collision is elastic. It seemed reasonable to conclude, on this basis, that the probability of radiative capture would always be small.[c] Such a conclusion, however, presupposes a point of view which regards the processes of scattering and capture as describable, in the first approximation, as single particle phenomena (cf. p. 47): in order to explain the experimental data at present under discussion we must clearly start from other assumptions. These will be introduced in the next section when the greatly enhanced capture probability for slow neutrons by certain nuclei may be taken as an experimentally established fact.

In this section we are concerned with the effects of high energy neutrons, and the last of these to be considered is that of non-capture excitation. The experiments are those of Lea[d] and of Fleischmann.[e] Lea examined twelve elements, from hydrogen to bismuth, using the neutrons from a polonium-

[a] The transformations ${}_{90}Th^{232}\ ({}_0n^1, {}_2He^4)\ {}_{88}Ra^{229}$, ${}_{90}Th^{232}\ ({}_0n^1, {}_1H^1)\ {}_{89}Ac^{232}$ and ${}_{92}U^{238}\ ({}_0n^1, {}_2He^4)\ {}_{90}Th^{235}$ have been suggested to account for certain features of the additional radioactivity artificially produced by bombarding thorium and uranium with neutrons: cf. Curie, v. Halban and Preiswerk, *J. Physique*, 6, 361, 1935; Hahn and Meitner, *Naturwiss.* 23, 320, 1935; Meitner and Hahn, *ibid.* 24, 158, 1936. Resonance would only be ruled out if it were found that the emitted particles (α particles and protons) possessed very high energies. This result is hardly likely.

[b] Plate III (*d*) is typical of many photographs showing disintegration with such a small amount of kinetic energy in the final products that escape of the disintegration particle "through the potential barrier"—either in conditions of resonance, or otherwise—must certainly be assumed.

[c] Cf. Massey and Mohr, *Nature*, 133, 211, 1934.

[d] Lea, *Proc. Roy. Soc.* 150, 637, 1935.

[e] Fleischmann, *Z. Physik*, 97, 242, 265, 1935. See also Arzimowitsch, Kurtchatow, Latychew and Chromow, *Phys. Z. Sow. Un.* 8, 472, 1935; Kikuchi, Aoki and Husimi, *Nature*, 137, 398, 1936.

beryllium source. By employing high pressure ionisation chambers in conjunction with tube counters he was able to show that the neutrons from the source (rather than the accompanying γ rays) were responsible for the production of quanta which were present, along with scattered neutrons, in the secondary radiation from each substance irradiated. The cross-section for production of this quantum radiation appeared to increase regularly (and slowly) with the atomic number of the target element. Obviously such experiments are very difficult to interpret satisfactorily, but definite quantitative comparisons were made which seemed to show that the process investigated was more efficient than the process of production of artificial radioactivity in similar circumstances. It could not, therefore, consist in simple capture and nothing else. Only one assumption remained possible: inelastic scattering of neutrons by heavy nuclei is a frequent occurrence. A rough estimate showed that perhaps one-half of the neutrons which suffer close collision with such a nucleus lose a considerable amount of energy in the collision. From the general viewpoint of the next section this result is not surprising.

§ 38. *Effects produced by neutrons of small energy and the nuclear model of Bohr.* If a nuclear transformation is to be produced by a projectile of very small energy two conditions, at least, must be fulfilled: the transformation process must be exothermic and the projectile must be capable of penetrating the nuclear structure, in spite of its lack of energy. The second condition excludes charged projectiles from consideration—exothermic processes may be possible (cf. §§ 36, 42), but charged particles of small kinetic energy are unable, in general, "to traverse the potential barrier surrounding the nucleus". Nevertheless, because many neutron-provoked transformations are also exothermic, there remains in them a wide range of possibility for processes in which the responsible particle is of very small energy. Transformations of the capture disintegration type, or pure capture processes, may occur in this way. In actual fact evidence for such transformations was obtained first in the case of pure capture processes; here, for sake of convenience, we shall reverse the chronological order and consider the emission of particles in disintegrations produced by slow neutrons, before discussing the more complicated subject of simple capture.

As already indicated, the emission of charged particles in neutron-provoked disintegrations becomes rapidly less probable as the atomic number of the bombarded element increases and the disintegration particle has to traverse a correspondingly high barrier of potential energy in its escape from the nucleus. Probably no further explanation than this is required of the fact that the only disintegrations so far observed with slow neutrons are those of the elements lithium, boron and nitrogen. The first observations were made by Chadwick and Goldhaber,[a] using an ionisation chamber and proportional amplifier,[b] the results with lithium being interpreted in terms of the reaction

$$_3\mathrm{Li}^6 + {}_0n^1 \rightarrow {}_2\mathrm{He}^4 + {}_1\mathrm{H}^3 \qquad \ldots\ldots(46).$$

Taylor and Goldhaber[c] then studied the disintegration of boron by the method of the photographic plate. As previously mentioned (p. 177), they showed that the corresponding transformation was (45), $_5\mathrm{B}^{10}\,({}_0n^1, {}_2\mathrm{He}^4)\,{}_3\mathrm{Li}^7$. Evidence for this transformation[d] has since been obtained from expansion chamber photographs.[e] The greater part of the information concerning the disintegration of nitrogen by slow neutrons has likewise been collected by this method.[f] It appears that, of the various transformations which are possible, it is (35), $_7\mathrm{N}^{14}\,({}_0n^1, {}_1\mathrm{H}^1)\,{}_6\mathrm{C}^{14}$, which is exothermic and takes place under these conditions.[g]

In respect of simple capture transformations the initial

[a] Chadwick and Goldhaber, *Nature*, 135, 65, 1935.

[b] See also Amaldi, *Nuovo Cim.* 12, 223, 1935; Chadwick and Goldhaber, *Proc. Camb. Phil. Soc.* 31, 612, 1935.

[c] Taylor (H. J.) and Goldhaber, *Nature*, 135, 341, 1935.

[d] Confirmatory evidence for (46) is similarly provided by work with the photographic plate, see Taylor and Dabholkar, *Proc. Phys. Soc.* 48, 285, 1936.

[e] Kurtchatow, Kurtchatow and Latychew, *Comptes rendus*, 200, 1199, 1935. See also Roaf, *Proc. Roy. Soc.* 153, 568, 1936.

[f] Bonner and Brubaker, *Phys. Rev.* 49, 223, 1936.

[g] Bonner and Brubaker, *Phys. Rev.* 49, 778, 1936.

discovery was due to Fermi.[a] It was found that a number of irregularities in previous work on artificially produced radioactive substances could be attributed to the fact that activation had sometimes been carried out in the vicinity of large amounts of water or paraffin wax. Since multiple scattering of neutrons in hydrogenous substances must rapidly reduce initial velocities to approximately "thermal" values,[b] it was realised that a marked variation of capture efficiency with velocity might be in question. Direct experiment quickly confirmed this suspicion: as the velocity of the neutrons was reduced[c] radiative capture obviously became much more frequent. Then it was found that the increase in efficiency varied in a most haphazard way from one element to another. This is a very pronounced—and very important—feature of the phenomenon. Under standard geometrical conditions an increase of fifty per cent. or of a hundredfold might be observed, depending upon the substance bombarded. Whilst these facts challenge explanation and must presently be discussed, we may notice, first of all, a result of some interest which depends directly upon a large increase in pure capture efficiency for small values of the neutron velocity. Utilising this effect it has become possible for the first time to obtain direct evidence of nuclear transformations of heavy elements, other than those which result in the production of unstable ("artificially radioactive") species. By employing relatively feeble neutron sources of polonium-beryllium, the γ radiation accompanying the capture of slow neutrons in cadmium and yttrium has been detected.[d] Since these elements do not acquire appreciable activity under such irradiation it must clearly be supposed that with them radiative capture results

[a] Fermi, Amaldi, Pontecorvo, Rasetti and Segrè, *La Ricerca Scientifica*, 5 (2), 282, 1934.

[b] Cf. Condon and Breit, *Phys. Rev.* 49, 229, 1936.

[c] Bombardment by slow neutrons is now always achieved by surrounding target and source (producing neutrons of high energy) with water, paraffin wax, or other hydrogenous material. The process of slowing down neutrons by this method has been studied by Bjerge and Westcott (*Proc. Roy. Soc.* 150, 709, 1935) and Westcott and Niewodniczánski (*Proc. Camb. Phil. Soc.* 31, 617, 1935), amongst many others.

[d] Amaldi, D'Agostino, Fermi, Pontecorvo, Rasetti and Segrè, *Proc. Roy. Soc.* 149, 522, 1935.

in the transformation of one stable isotope into another.[a] No doubt this effect occurs quite frequently;[b] usually, however, it is masked by the simultaneous production of an unstable species through the transformation of another isotope—or escapes detection with polonium-beryllium sources[c] because of its relatively small increase in probability with slow neutrons.

We return now to the evidence more easily obtained by investigating the production of artificial radioactivity (or merely by studying the absorption and scattering of slow neutrons, using the disintegration of lithium or boron for purposes of detection). Because this evidence is accumulating so rapidly no attempt can be made to follow individual investigations or record particular results. It must be sufficient to indicate broadly the general aim of the experiments and to point to any significant conclusions which may be drawn from the data which they provide. Generally this aim is to deduce precisely the variation of capture efficiency with neutron energy for each element. The results, though hopelessly incomplete in this respect, already clearly indicate that, even within the range of thermal energies and somewhat above, very frequently the variation in capture efficiency is not monotonic. Fairly sharp "absorption bands" occur in the energy spectrum.[d] This fact, the apparent independence of the whole phenomenon of atomic number (p. 182), and the result that high probability of radiative capture and small probability of elastic scattering often characterise a single element, must be the basis of our theoretical discussion.

[a] With yttrium this would seem rather a doubtful conclusion, on general grounds: the atomic number is odd and evidence at present suggests that yttrium is a "pure" element. In any case, two stable isotopes for which the mass numbers differ by one unit are hardly to be expected (cf. p. 76). Perhaps an activity of very long period is in question. (This has now been found: cf. Hevesy and Levi, *Nature*, 137, 185, 1936.)

[b] Cf. Rasetti, *Z. Physik*, 97, 64, 1935. See also p. 208.

[c] Intense γ rays from the source make similar experiments very much more difficult with the stronger radon-beryllium sources of neutrons.

[d] An absorption band at, say, 3 electron volts energy with a "half-breadth" of 0·2 electron volt (the case of capture by one of the isotopes of silver) would here be referred to as fairly sharp. When, however, this half-breadth is regarded as belonging equally to a γ ray line at roughly 10^7 electron volts energy in the spectrum of the capture radiation, the latter may justly be considered as exceedingly narrow.

It may be said, at once, that the last mentioned result alone is decisive against the earlier attempts at interpretation. Of these the calculations of Bethe[a] may be taken as representative; they predict a variation in the cross-section for elastic scattering of neutrons which is altogether parallel to that of the cross-section for radiative capture. Two suggestions to remedy this defect in the theory are at present in the field: Bohr[b] has taken as starting-point the complete abandonment of the one-particle approximation in the discussion of nuclear phenomena and Breit and Wigner,[c] whilst attempting a treatment which follows more closely the regular methods of wave mechanics, are also forced, in all considerations of energy levels, to refer these to the system (neutron + nucleus) as a single unit. In this respect the two standpoints have much in common. Bohr's theory, in its particular application to slow neutron effects, makes use of the fact that the nuclear aggregate (neutron + nucleus) under these conditions (i.e. before the radiation of energy) will generally have energy in excess of the normal by nearly 10^7 electron volts (cf. Fig. 4 (*a*)). We have to enquire, therefore, what will be the nature of the stationary states of this system for approximately this energy of excitation. If it be supposed that additional energy is taken up by some quantised collective motion of all the nuclear particles together, then clearly the greater the amount of energy so absorbed the greater also the multiplicity of nuclear states in which it is capable of being retained. Nuclear energy levels rapidly become more and more closely spaced the greater the energy of excitation.[d] From this point of view there need be no contradiction between the γ ray data, on the one hand—exhibiting widely spaced energy levels in heavy nuclei for 1 or 2×10^6 electron volts energy of excitation—and the fact, demonstrated by the occasional occurrence of resonance effects in slow neutron capture, on the other, that at the higher excitations (say 10^7 electron volts excess energy) the probability that an

[a] Bethe, *Phys. Rev.* 47, 747, 1935. [b] Bohr, *Nature*, 137, 344, 1936.
[c] Breit and Wigner, *Phys. Rev.* 49, 519, 1936.
[d] Cf. Bethe, *Phys. Rev.* 50, 332, 1936.

energy level shall be found in any arbitrarily chosen small interval of energy (say 0 to 10 electron volts for the incident neutron) is not excessively remote. The haphazard incidence —within the system of the elements—of greatly enhanced capture at low velocities, and the evidence of sharp absorption bands, now become mutually explanatory facts. It remains to be seen how the newly proposed model is also capable of explaining the relatively small extent of the elastic scattering of slow neutrons in strongly absorbing materials. The argument is roughly as follows. Scattering is regarded as normally a dual process; a neutron enters the nuclear structure and, after a finite time, a neutron is again emitted. Clearly such re-emission can only take place when sufficient energy is once more concentrated in a single nuclear particle; when the incident neutron is of small energy this may take a very long time. In such circumstances the emission of radiation, resulting in permanent capture, is a much more probable occurrence. Now, the only suspicion which remains is that the explanation provides almost too stringent a prohibition! Apart from this suspicion, however, very little criticism is possible. In respect of the radiative capture of neutrons by nuclei, obviously the ideas of Bohr provide a most successful basis of interpretation. Whether that success will be extended to considerations of the disintegration effects produced by high energy neutrons—and by charged particles—cannot be apprised at this stage.[a] There appears to be much less necessity, with disintegration phenomena, for regarding the nuclear change energetically otherwise than as a single event.

§ 39. *Collected results.*

(*a*) Non-capture excitation.

See p. 179.

(*b*) Non-capture disintegration.

See p. 171.

Recently the transformation

$${}_{92}\mathrm{U}^{238}\ (+{}_0n^1) \rightarrow {}_{92}\mathrm{U}^{237} + {}_0n^1\ (+{}_0n^1)$$

[a] Cf. Gamow, *Phys. Rev.* 49, 946, 1936.

has been postulated to explain certain features of the artificial radioactivity induced in uranium under neutron bombardment.[a]

(*c*) Radiative capture.

Neutrons are known to be captured by the following nuclei. Undoubtedly this process is of more universal occurrence than is indicated by this list. H^1, F^{19}, Na^{23}, Mg^{26}, Al^{27}, Si^{30}, P^{31}, Cl^{37}, A^{40}, K^{41}, Ca^{44}, $\underline{V^{51}}$, $\underline{Mn^{55}}$, $\underline{Cu^{63,65}}$, Ga^{69}, As^{75}, $\underline{Br^{79,81}}$, Y^{89}, $\underline{Rh^{103}}$, $\underline{Ag^{107,109}}$, $\underline{In^{113,115}}$, I^{127}, Cs^{133}, La^{139}, Pr^{141}, Tb^{159}, Lu^{175}, Hf^{180}, $\underline{W^{186}}$, Ir^{193}, Au^{197}, Hg^{204}, $Tl^{203,205}$, Th^{232}, U^{238}.

When the efficiency of capture is markedly greater for neutrons of small energy than for high energy neutrons the nuclear symbol is here underlined.

(*d*) Capture disintegration.

(i) Disintegrations of the type ${}_ZX^A\ ({}_0n^1, {}_2\mathrm{He}^4)\ {}_{Z-2}Y^{A-3}$.

Table 18

Target nucleus	Q e.v. $\times 10^{-6}$	Product nucleus
${}_3Li^6$	4·3	${}_1H^3$
${}_5B^{10}$	2·25	${}_3Li^7$
${}_6C^{12}$	(−7)	${}_4Be^9$
${}_7N^{14}$	(0 to −10)	${}_5B^{11}$
${}_8O^{16}$	(−2 to −8)	${}_6C^{13}$
${}_9F^{19}$	(−2 to −6)	${}_7N^{16}$
${}_{10}Ne^{20}$	(−0·7, −5·3)	${}_8O^{17}$

In neutron-provoked disintegrations obviously the energy release may be determined with greater accuracy when the captured neutron is itself of small energy. Data obtained with high energy neutrons are therefore bracketed in Table 18. Also (Q values not known) Be^9,[b] Mg^{26}, Al^{27}, P^{31}, Cl^{35}, Sc^{45}, Mn^{55}, Co^{59}, Th^{232} and U^{238} undergo disintegration of this type.

(ii) Disintegrations of the type ${}_ZX^A\ ({}_0n^1, {}_1\mathrm{H}^1)\ {}_{Z-1}Y^A$.

[a] Meitner and Hahn, *Naturwiss.* 24, 158, 1936.

[b] Bjerge, *Nature*, 137, 865; 138, 400, 1936. Probably ${}_4Be^9\ ({}_0n^1, {}_2He^4)\ {}_2He^6$ explains the artificial radioactivity found when beryllium is bombarded by neutrons.

$${}_7N^{14}\ ({}_0n^1,\ {}_1H^1)\ {}_6C^{14}; \qquad Q = 0{\cdot}6 \times 10^6 \text{ electron volts.}$$

Also (Q values not known) F^{19}, Na^{23}, Mg^{24}, Al^{27}, Si^{28}, P^{31}, S^{32}, Cl^{35}, Ca^{42}, Cr^{52}, Fe^{56}, $Zn^{64,\,66}$ and Th^{232} undergo disintegration of this type.

(iii) Three-particle disintegrations of the type

$${}_ZX^A\ ({}_0n^1,\ 2\,{}_2He^4)\ {}_{Z-4}Y^{A-7}.$$

B^{10}, C^{12}, N^{14} undergo disintegration of this type.

CHAPTER XII

TRANSFORMATIONS PRODUCED BY ACCELERATED PARTICLES

§ 40. *Transformations produced by fast-moving protons.* It has already been remarked (p. 27) that, with artificially accelerated particles, the most significant results concerning nuclear transformation are to be expected with the simplest (lightest) particles. Hitherto, in actual fact, the important experimental investigations have been concerned entirely with the effects produced by fast-moving protons and deuterons. In this section we shall deal with the transformations resulting from proton bombardment, only, but it will be well to preface our discussion by reference to the negative results of experiments in which electrons have been used. In many ways it is simpler to produce intense beams of high energy electrons than of any other particles—but there are difficulties in mere intensity (heating effects, X radiation, etc.) which have not always been appreciated. Thus, in the past, transmutation has been claimed on a number of occasions, but further research has invariably negatived the earlier result. Now, the technique of investigation by the method of artificial radioactivity (p. 177) has considerably increased the power of the investigator. It is satisfactory, therefore, that the latest exhaustive survey by this method[a] has failed to reveal the slightest evidence of any nuclear change initiated by electron capture. Livingood and Snell found no sign of artificial radioactivity after bombarding many elements with electrons of $0{\cdot}8 \times 10^6$ electron volts energy under the most favourable conditions.[b] This is satisfactory chiefly because of the relevance of these experiments to the theory of β disintegration: the non-occurrence of electron-provoked trans-

[a] Livingood and Snell, *Phys. Rev.* 48, 851, 1935.

[b] Electron currents of about 100 microamperes were used.

formations[a] provides considerable support for the view that β processes in general are essentially three particle phenomena; the reverse of β particle emission is not merely electron capture, only, it involves the simultaneous capture of a neutrino or ejection of an anti-neutrino. Hitherto no experiments have been carried out in conditions in which there is considerable probability of the simultaneous entry of an electron and a neutrino into a single target nucleus: such conditions will clearly be very difficult of realisation.

Returning to proton-provoked transformations we may notice two facts, in particular. The first case of disintegration produced by artificially accelerated particles was of this type—and, in all, these transformations have turned out to be much less numerous and varied than those observed under deuteron bombardment (§ 41). In 1932 Cockcroft and Walton[b] observed the scintillations of what appeared to be α particles of about 8·4 cm. range (in standard air) when a target containing lithium was bombarded by protons of $0{\cdot}5 \times 10^6$ electron volts energy. Their interpretation of the transformation according to the scheme

$$_3\mathrm{Li}^7\ (_1\mathrm{H}^1,\ _2\mathrm{He}^4)\ _2\mathrm{He}^4 \qquad \text{......(47)}$$

has since been verified in every particular: the assumption that the heavier lithium isotope was concerned was substantiated by experiments with samples containing the two isotopes separately,[c] whilst the conclusion that α particles should be observed in pairs (of nearly the same energy and approximately oppositely directed)—rather than singly—was tested and proved correct by experiments with the expansion chamber.[d] An accurate determination of the amount of kinetic energy released in the transformation has given[e] $Q = 17{\cdot}06 \pm 0{\cdot}06 \times 10^6$ electron volts. It may be re-

[a] See, however, Tanaka (*Phys. Rev.* 48, 916, 1935). Further negative results have recently been reported by Lewis and Burcham, *Proc. Camb. Phil. Soc.* 32, 503, 1936.

[b] Cockcroft and Walton, *Proc. Roy. Soc.* 137, 229, 1932.

[c] Oliphant, Shire and Crowther, *Nature*, 133, 377, 1934; *Proc. Roy. Soc.* 146, 922, 1934.

[d] Dee and Walton, *Proc. Roy. Soc.* 141, 733, 1933.

[e] Oliphant, Kempton and Rutherford, *Proc. Roy. Soc.* 149, 406, 1935.

marked that this large release of energy is clearly to be explained by the condensation of nuclear constituent particles with the formation of a new α particle (p. 62)—and the non-occurrence of paired α particles of less than the maximum energy by the assumption that no γ radiation is emitted, either before or after the emission of particles (cf. p. 193).

We have discussed the nature of the reaction and the balance of energy: the final problem (p. 154), concerning the dependence of disintegration yield upon the initial energy of the protons, has been investigated by many workers. The original results of Cockcroft and Walton were here extended, first towards higher energies by Henderson[a] (and later by Hafstad and Tuve),[b] and then in the low energy region by Oliphant and Rutherford[c] and Jackowlew.[d] Recently, Herb, Parkinson and Kerst[e] have obtained more accurate data for protons of intermediate energies. A discussion of the results then available was given by Cockcroft in 1934;[f] a more detailed treatment, including the latest experimental material, has now appeared.[g] Many points of interest are raised, but perhaps the one which most needs to be remembered is the warning that "the height of the nuclear potential barrier against protons" cannot be deduced by any naïve inspection of the excitation curve. The potential energy of a proton inside the nucleus—as well as the variation of this quantity with distance outside—is important in determining the energy dependence of disintegration yield.

Under proton bombardment disintegrations of the general type

$$_{Z}X^{A}\ ({}_{1}\mathrm{H}^{1},\ {}_{2}\mathrm{He}^{4})\ {}_{Z-1}Y^{A-3}$$

have been established with a number of elements in addition to lithium.[h] The relevant data will be found in § 42 and, except for one case, need no further discussion. The exception,

[a] Henderson (M. C.), *Phys. Rev.* 43, 98, 1933.
[b] Hafstad and Tuve, *Phys. Rev.* 48, 306, 1935.
[c] Oliphant and Rutherford, *Proc. Roy. Soc.* 141, 259, 1933.
[d] Jackowlew, *Z. Physik*, 93, 644, 1935; cf. Doolittle, *Phys. Rev.* 49, 779, 1936.
[e] Herb, Parkinson and Kerst, *Phys. Rev.* 48, 118, 1935.
[f] Cockcroft, *Int. Conf. Phys.* 1, 112, 1935 (§ 6).
[g] Ostrofsky, Breit and Johnson, *Phys. Rev.* 49, 22, 1936; Ostrofsky, Bleick and Breit, *ibid.* 49, 352, 1936.
[h] Also, in one instance, a disintegration of the type $_{Z}X^{A}\ ({}_{1}\mathrm{H}^{1},\ {}_{1}\mathrm{H}^{2})\ {}_{Z}X^{A-1}$ has been shown to occur.

here, is the case of boron. Soon after the first discovery of the disintegration of boron under proton bombardment it became clear[a] that this transformation showed unusual features; the α particles which were observed in a given direction were not homogeneous in velocity as would be expected from a two particle disintegration with discrete values of Q. It was suggested that, in fact, three particles were involved. The simplest supposition was that each of these particles was an α particle:

$$_5B^{11} + {}_1H^1 \rightarrow 3\,{}_2He^4 \qquad \ldots\ldots(48).$$

Then Kirchner and Neuert[b] found evidence for a group of mono-energetic particles of small intensity—and this observation was subsequently verified.[c] Originally Kirchner[d] had assumed these particles to be the α particles produced in the transformation

$$_5B^{11} + {}_1H^1 \rightarrow {}_4Be^8 + {}_2He^4 \qquad \ldots\ldots(49);$$

this identification, also, became generally accepted. There remained, however, considerable uncertainty—and some difficulties of interpretation—regarding the connection, if any, between the two-particle and three-particle modes of disintegration. These difficulties have now been removed by the experiments of Dee and Gilbert.[e] Dee and Gilbert explain both the angular distribution and the distribution of energy (Fig. 20) amongst the products of disintegration in the following terms. The transformation is initially a two-particle transformation in all cases, but in a small fraction, only, is the final nucleus (Be^8) formed in the ground state: usually this nucleus is produced with about 3×10^6 electron volts energy of excitation. This state, however, is unstable in respect of particle disintegration[f] and its extremely short lifetime ($\sim 10^{-21}$ sec.) has two important consequences. In the first place its energy content is ill-defined (Q values for the transformations which leave the residual nucleus in this excited state appear to be distributed over a range of energy roughly

[a] Cockcroft and Walton, *Nature*, 131, 23, 1933.
[b] Kirchner and Neuert, *Phys. Z.* 35, 292, 1934.
[c] Oliphant, Kempton and Rutherford, *Proc. Roy. Soc.* 150, 241, 1935.
[d] Kirchner, *Naturwiss*, 22, 480, 1934.
[e] Dee and Gilbert, *Proc. Roy. Soc.* 154, 279, 1936.
[f] If the ground state is stable it appears that the energy of binding is very small indeed: more probably the ground state is unstable, also.

2×10^6 electron volts in extent) and, secondly, the probability of de-excitation by particle emission is much greater than that of de-excitation by radiation. Thus, almost immediately, the excited nucleus disintegrates (without any directional bias) into two α particles. Broadly speaking, of the three α particles resulting from the capture of a proton by a particular nucleus (according to (49)), one—the α particle first emitted—has approximately the same energy in all cases; the energies

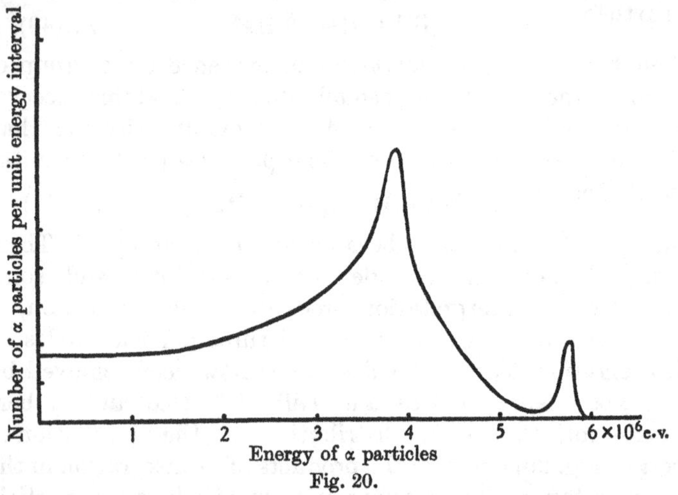

Fig. 20.

of the other two are determined by the circumstances of the disintegration of the recoiling residue. It is interesting to remark that five other examples[a] of three particle disintegrations are known in which two of the particles set free are α particles. It is possible that some of these transformations take place in two stages as indicated above. The nucleus Be^8 may well be unique amongst the lighter species in being so near the limit of stability (in respect of heavy particle disintegration) in its ground state; we may suppose that, in these cases, also, it is the (unstable) residue first formed.[b]

[a] Cf. pp. 187, 200.

[b] It is possible, alternatively, that the disintegration of carbon by neutrons, (43), proceeds initially according to the scheme ${}_6C^{12}$ (${}_0n^1$, ${}_2He^4$) ${}_4Be^9$ and that Be^9 (in the ground state) is almost unstable in respect of the transformation

$${}_4Be^9 \rightarrow {}_2He^5 + {}_2He^4.$$

It has been stated (p. 190) that there is no evidence from the analysis of the corpuscular products of transformation for nuclear excitation following the disintegration of lithium by protons. A similar statement may be taken as true of the disintegration of fluorine in the corresponding mode[a]

$$_9F^{19}\ (_1H^1,\ _2He^4)\ _8O^{16}.$$

Nevertheless, at quite an early stage, it was found that γ radiation was produced, in addition to particles, with these two elements.[b] Recently, similar observations of γ radiation, involving essentially the same difficulties of interpretation, were reported when beryllium and boron[c] were bombarded. It is very likely that the correct explanation of all these effects is to be found along the lines suggested by a previous interpretation of the artificial radioactivity produced in carbon (and boron) under proton bombardment.[d] Simple capture of the proton in the transformations

$$_6C^{12} + {_1H^1} \rightarrow {_7N^{13}} \qquad \text{......(50)},$$

$$_5B^{10} + {_1H^1} \rightarrow {_6C^{11}} \qquad \text{......(51)}$$

was advanced as the only reasonable hypothesis—particularly after it was found that the detailed phenomena of recoil of the radioactive nucleus were completely in accord with this suggestion. Simple capture in the other cases mentioned must result in the production of the stable nuclei Be^8, Ne^{20}, B^{10} and C^{12}, respectively. From the theoretical point of view the process of radiative capture of a proton has been discussed by Wilson[e] and, more recently, by Breit and Yost.[f] In the later theoretical treatment[g] the question of resonance occupies a prominent position. From the experimental side, also, characteristic resonance effects have been observed.

[a] Henderson (M. C.), Livingston and Lawrence, *Phys. Rev.* 46, 38, 1934.

[b] Crane, Delsasso, Fowler and Lauritsen, *Phys. Rev.* 46, 531, 1934; *ibid.* 48, 125, 1935.

[c] Crane, Delsasso, Fowler and Lauritsen, *Phys. Rev.* 47, 782; *ibid.* 48, 102, 1935.

[d] Cockcroft, Gilbert and Walton, *Proc. Roy. Soc.* 148, 225, 1935.

[e] Wilson (A. H.), *Mon. Not. R.A.S.* 91, 283, 1931.

[f] Breit and Yost, *Phys. Rev.* 48, 203, 1935.

[g] No doubt this process, also, is capable of inclusion in the alternative scheme of explanation proposed by Bohr (§ 38).

Hafstad and Tuve[a] obtained the results given in Fig. 21 (A) in respect of the γ radiation produced by bombarding lithium with protons,[b] the excitation function for the two-particle process (47) being included in the figure (B) for sake

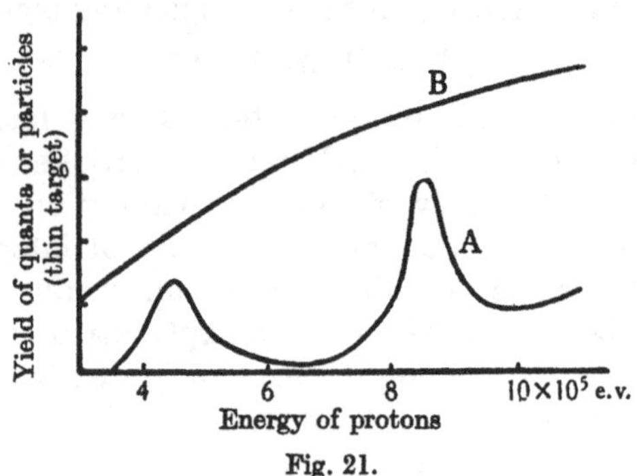

Fig. 21.

of comparison. Considering the curves A and B together, it is evident that resonance here does not represent a condition of particularly easy entry of the incident particle into the nuclear structure (otherwise indications of resonance would occur in curve B, also); it must represent a condition particularly favourable to the radiation of the excess energy which has to be got rid of before the process of capture is possible.

§ 41. *Transformations produced by fast-moving deuterons.* Within a very short time of the discovery of the hydrogen isotope of mass number two the corresponding positive ions—deuterons—were employed in experiments on nuclear transformation by Lewis, Livingston and Lawrence.[c] It was shown that long range α particles were obtained when a target of lithium was bombarded by these projectiles. Since that time most of the elements have been subjected to a similar bombardment and a surprisingly large number of nuclear transformations has been set in evidence. Neutrons, protons, H^3 nuclei and α particles have been shown to be emitted, as disintegration products of smaller mass, in one disintegration

[a] Hafstad and Tuve, *Phys. Rev.* 47, 506; *ibid.* 48, 306, 1935.

[b] It is now known that only the peak at 440 e.kV. in A refers to the capture of protons by lithium nuclei.

[c] Lewis (G. N.), Livingston and Lawrence, *Phys. Rev.* 44, 55, 1933.

process or another. Initially, interpretation of the results with lithium was complicated by the fact that lithium nuclei of two types were present, and both were obviously affected. Now, however, work with the separated isotopes[a]—and a more detailed examination of the effects produced with ordinary targets—has shown that the following transformations take place:

$$_3Li^6 + {}_1H^2 \rightarrow 2\,{}_2He^4 \qquad ......(52),$$

$$_3Li^6 + {}_1H^2 \rightarrow {}_3Li^7 + {}_1H^1 \qquad ......(53),$$

$$_3Li^7 + {}_1H^2 \rightarrow {}_4Be^8 + {}_0n^1 \qquad ......(54)^b,$$

$$_3Li^7 + {}_1H^2 \rightarrow 2\,{}_2He^4 + {}_0n^1 \qquad ......(55),$$

$$_3Li^7 + {}_1H^2 \rightarrow {}_3Li^8 + {}_1H^1 \qquad ......(56)^c.$$

The α particles of longest range, under given conditions, are those produced in process (52). With beryllium and boron just about the same wide range of processes has been established, and with other light elements disintegration products, both stable and radioactive, have been found in great variety. The relevant information is given in tabular form in § 42. Here we must discuss these transformations from two different viewpoints. We are concerned with the energy changes which they involve—and, as on previous occasions, we shall have to enquire how the probability of disintegration in each case varies with the energy of the deuterons. The problem of the energy changes may be further subdivided; it will be interesting to examine the different values of Q characteristic of the formation of various stable product nuclei, and it is important to make some attempt to relate the energies of formation of alternative stable and unstable nuclei with the maximum energy of the β particles subsequently emitted by the latter (cf. Tables 12 and 13).

The different Q values in the transformation

$$_5B^{10}\,({}_1H^2,\,{}_1H^1)\,{}_5B^{11}$$

have already been discussed (p. 173) in connection with the disintegration of nitrogen by neutrons. We have seen that no

[a] See p. 189; also Kempton, Browne and Maasdorp, *Proc. Roy. Soc.* in the press.

[b] Bonner and Brubaker, *Phys. Rev.* 48, 742, 1935.

[c] Delsasso, Fowler and Lauritsen, *Phys. Rev.* 48, 848, 1935.

inconsistencies arise if the view be adopted that differences in Q represent energies of excitation of the product nucleus—and if it be assumed that a given nucleus is capable of excitation to the same states[a] when produced in quite different transformations. The various transformations in which the nucleus C^{12} is produced provide further opportunity for testing this assumption. With α particles this nucleus results from the bombardment of beryllium (p. 162); with deuterons it may be produced from boron (B^{11}) and nitrogen (N^{14}), respectively. With deuterons and boron Q values have not been investigated in detail, but with nitrogen, when α particles are emitted,

$$_7N^{14} + _1H^2 \rightarrow _6C^{12} + _2He^4 \qquad \ldots\ldots(57),$$

the values 13·2 and $8{\cdot}9 \times 10^6$ electron volts have been clearly established. The difference between these two energies, $4{\cdot}3 \times 10^6$ electron volts, agrees very closely indeed with the quantum energy of one component in the spectrum of the γ radiation from beryllium bombarded by α particles (footnote, p. 162).[b] All this is strong evidence in favour of our initial assumptions. It is the more conclusive in that, if it is correct to discuss any features of capture disintegration in terms of an unstable nuclear aggregate first formed, then, since this intermediate product is not the same in the transformations (57) and $_4Be^9$ ($_2He^4$, $_0n^1$) $_6C^{12}$, it is clearly of no avail to interpret any features which these disintegrations have in common with the help of this idea.

The alternative transformations most suited to the discussion of the energy balance when an artificially radioactive product is involved are the disintegrations

$$_6C^{12} + _1H^2 \rightarrow _6C^{13} + _1H^1 \qquad \ldots\ldots(58)$$

and

$$_6C^{12} + _1H^2 \rightarrow _7N^{13} + _0n^1 \qquad \ldots\ldots(59).$$

[a] Cf. Feather, *Proc. Roy. Soc.* 142, 689, 1933.

[b] A γ radiation of quantum energy $4{\cdot}0 \times 10^6$ e.v. has been reported (Crane, Delsasso, Fowler and Lauritsen, *Phys. Rev.* 48, 100, 1935) from nitrogen bombarded by deuterons. It is probable that the component of $5{\cdot}3 \times 10^6$ e.v. energy in the same spectrum is to be related to the energy difference $(8{\cdot}5 - 3{\cdot}25) \times 10^6$ e.v. found by examining the protons from $_7N^{14}$ ($_1H^2$, $_1H^1$) $_7N^{15}$.

Of these the latter is followed by the spontaneous emission of positrons (∂),

$$_{7}N^{13} \rightarrow {}_{6}C^{13} + \partial \qquad \text{......(60).}$$

We wish to discuss these reactions from the point of view of the current (neutrino) theory of β emission. If Q_1, Q_2 represent (in mass units) the amounts of energy released in (58) and (59), respectively, and E_0, similarly, the maximum energy of the positrons in (60), clearly these quantities and ω, the mass of the neutrino, are connected by the relation

$$N^{13} - C^{13} = H^1 - n^1 + Q_1 - Q_2 = 2e + \omega + E_0 \qquad \text{......(61),}$$

when atomic masses are employed (cf. footnote, p. 79). The experimental values are as follows:

$$H^1 - n^1 = -0{\cdot}0009 \pm 0{\cdot}0001, \quad Q_1 = 0{\cdot}00284,$$
$$2e = 0{\cdot}0011, \qquad Q_2 = -0{\cdot}00040,$$
$$E_0 = 0{\cdot}00134.$$

Thus $\omega = -0{\cdot}0001 \pm 0{\cdot}0001$: within the limits of accuracy of the data available the rest mass of the neutrino appears to be zero. It will be remembered that a similar conclusion has previously been reached on quite other evidence (p. 135); the agreement, therefore, is highly satisfactory. It will be most important to see whether it remains equally good when sufficiently accurate data are obtained in a larger number of cases.

We pass now to a consideration of the dependence of disintegration yield on deuteron energy. A general survey of the list of elements which suffer disintegration under deuteron bombardment—and of the types of disintegration which occur with each—reveals a very striking fact (cf. § 42). Disintegration, at least when it results in the emission of a proton (or a neutron), is not confined to the elements of low atomic number; in spite of the repulsion which a charged particle must experience in approaching a heavy nucleus (and of the difficulty of such a particle leaving the nucleus, if it is situated inside) these disintegrations take place. It was important, therefore, to investigate the energy dependence of disintegration yield for all transformations of the type

$_{Z}X^{A}$ ($_{1}H^{2}$, $_{1}H^{1}$) $_{Z}X^{A+1}$, in order to see whether some explanation of this fact might possibly be found. Utilising the radioactivity of the unstable products formed in a number of such transformations, Lawrence, McMillan and Thornton[a] carried out the investigation in respect of the target nuclei Na^{23}, Al^{27}, Si^{30} and $Cu^{63, 65}$. Later, Henderson[b] made a direct comparison of the excitation functions for the transformations

$$_{12}Mg^{26} + _{1}H^{2} \rightarrow _{12}Mg^{27} + _{1}H^{1} \qquad \text{......(62)}$$

and

$$_{12}Mg^{26} + _{1}H^{2} \rightarrow _{11}Na^{24} + _{2}He^{4} \qquad \text{......(63).}$$

The general result of all these experiments[c] is that the probability of disintegration with the emission of protons is much less dependent upon the energy of the deuterons than would be expected on any theory which treats of the penetration of a nuclear potential barrier in the usual manner. By contrast, the excitation function for the transformation (63) was of the standard type. An interesting suggestion has been made by Oppenheimer and Phillips[d] in order to explain these results. These authors assume that the disintegrations in question take place essentially by capture of a neutron from the incident deuteron without actual penetration of this particle into the nuclear structure.[e] The probability of capture under given conditions will clearly depend upon the energy of binding of the deuteron. It is favourable to the hypothesis of Oppenheimer and Phillips that the calculations based upon it fit satisfactorily with the experimental excitation curves only when a reasonable value is adopted for this energy of binding. In view of this numerical success, disintegrations of the type $_{Z}X^{A}$ ($_{1}H^{2}$, $_{1}H^{1}$) $_{Z}X^{A+1}$ may plausibly be referred to as rejection transformations—in that the neutron is captured and the proton rejected.

In this section and the last we have dealt with the transformations produced by accelerated hydrogen ions; protons and deuterons. It merely remains to be recorded, in conclu-

[a] Lawrence, McMillan and Thornton, *Phys. Rev.* 48, 493, 1935.
[b] Henderson (M. C.), *Phys. Rev.* 48, 855, 1935.
[c] Cf. Snell, *Phys. Rev.* 49, 555, 1936; Cork and Lawrence, *ibid.* 49, 778, 1936.
[d] Oppenheimer (J. R.) and Phillips, *Phys. Rev.* 48, 500, 1935.
[e] Transformation (63), of course, is impossible without penetration.

sion, that accelerated helium ions have, in one instance, been similarly employed[a]—without, however, approaching the efficiency obtained by the use of α particles from natural sources—and that high voltage experiments have also been carried out with lithium ions. Whatever future there may be for such experiments, the results, up to date,[b] add nothing to what has been discovered by other means.

§ 42. *Collected results.*

(i) Transformations produced by protons.

(*a*) Radiative capture.

Li^7, Be^9, B^{10}, B^{11}, C^{12}, F^{19}.

(*b*) Capture disintegration.

Disintegrations of the type $_ZX^A\ ({}_1H^1, {}_2He^4)\ {}_{Z-1}Y^{A-3}$.

Table 19

Target nucleus	$_3Li^6$	$_3Li^7$	$_4Be^9$	$_5B^{11}$	$_9F^{19}$
Q: e.v. $\times 10^{-6}$	3·6	17·06	2·05	8·5	8·2
Product nucleus	$_2He^3$	$_2He^4$	$_3Li^6$	$_4Be^8$	$_8O^{16}$

Disintegration of the type $_ZX^A\ ({}_1H^1, {}_1H^2)\ {}_ZX^{A-1}$.

$_4Be^9\ ({}_1H^1, {}_1H^2)\ {}_4Be^8$; $Q = 0{\cdot}5 \times 10^6$ electron volts.

Three-particle disintegration

$_5B^{11} + {}_1H^1 \rightarrow 3\,{}_2He^4$; $Q = 8{\cdot}7 \times 10^6$ electron volts.

(ii) Transformations produced by deuterons.

Capture disintegration.

Disintegrations of the type $_ZX^A\ ({}_1H^2, {}_0n^1)\ {}_{Z+1}Y^{A+1}$.

Table 20

Target nucleus	$_1H^2$	$_3Li^7$	$_4Be^9$	$_5B^{11}$	$_6C^{12}$	$_8O^{16}$
Q: e.v. $\times 10^{-6}$	3·2	14·3	4·25	13·5	−0·4	−1·8
Product nucleus	$_2He^3$	$_4Be^8$	$_5B^{10}$	$_6C^{12}$	$_7N^{13}$	$_9F^{17}$

Also (Q values not known) B^{10}, N^{14}, Na^{23} and Al^{27} undergo disintegration of this type.

[a] Crane, Lauritsen and Soltan, *Phys. Rev.* 45, 507, 1934.

[b] Zeleny, Brasefield, Bock and Pollard, *Phys. Rev.* 46, 318, 1934; Thornton and Kinsey, *ibid.* 46, 324, 1934; Whitmer and Pool, *ibid.* 47, 795, 1935; Petuhow, Sinelnikow and Walther, *Phys. Z. Sow. Un.* 8, 212, 1935; Kinsey, *Phys. Rev.* 50, 386, 1936.

Disintegrations of the type ${}_ZX^A$ (${}_1H^2$, ${}_1H^1$) ${}_ZX^{A+1}$.

Table 21

Target nucleus Q: e.v. × 10^{-6} Product nucleus	${}_1H^2$ 3·97 ${}_1H^3$	${}_3Li^6$ 5·0 ${}_3Li^7$	${}_3Li^7$ 4·3 ${}_3Li^8$	${}_4Be^9$ 4·7 ${}_4Be^{10}$	${}_5B^{10}$ 9·1, 7·0, 4·6 ${}_5B^{11}$	${}_6C^{12}$ 2·65 ${}_6C^{13}$

${}_7N^{14}$ 8·5, 3·25 ${}_7N^{15}$	${}_8O^{16}$ 1·9 ${}_8O^{17}$	${}_9F^{19}$ 1·7 ${}_9F^{20}$	${}_{11}Na^{23}$ 4·5, 1·5 ${}_{11}Na^{24}$	${}_{13}Al^{27}$ 5·3, 4·7, 2·8, 1·9, 0·5 ${}_{13}Al^{28}$	${}_{17}Cl^{37}$ 3·5 ${}_{17}Cl^{38}$

Also (Q values not known) B^{11}, N^{15}, Mg^{26}, Si^{30}, P^{31}, A^{40}, K^{41}, Cu, Zn, As^{75} and Pt undergo disintegration of this type.

Disintegration of the type ${}_ZX^A$ (${}_1H^2$, ${}_1H^3$) ${}_ZX^{A-1}$.

$${}_4Be^9\ ({}_1H^2, {}_1H^3)\ {}_4Be^8; \quad Q = 4{\cdot}5 \times 10^6 \text{ electron volts.}$$

Disintegrations of the type ${}_ZX^A$ (${}_1H^2$, ${}_2He^4$) ${}_{Z-1}Y^{A-2}$.

Table 22

Target nucleus Q: e.v. × 10^{-6} Product nucleus	${}_3Li^6$ 22·06 ${}_2He^4$	${}_4Be^9$ 7·2 ${}_3Li^7$	${}_5B^{10}$ 17·5 ${}_4Be^8$	${}_5B^{11}$ 8·1 ${}_4Be^9$	${}_6C^{13}$ 5·1 ${}_5B^{11}$	${}_7N^{14}$ 13·2, 8·9 ${}_6C^{12}$

${}_8O^{16}$ 2·95 ${}_7N^{14}$	${}_{11}Na^{23}$ 6·9 ${}_{10}Ne^{21}$	${}_{12}Mg^{26}$ 8 ${}_{11}Na^{24}$	${}_{13}Al^{27}$ 6·6 ${}_{12}Mg^{25}$

Three and four particle disintegrations:

$${}_3Li^7 + {}_1H^2 \rightarrow 2\,{}_2He^4 + {}_0n^1,$$
$${}_5B^{10} + {}_1H^2 \rightarrow 3\,{}_2He^4,$$
$${}_5B^{11} + {}_1H^2 \rightarrow 3\,{}_2He^4 + {}_0n^1,$$
$${}_7N^{14} + {}_1H^2 \rightarrow 4\,{}_2He^4.$$

CHAPTER XIII

EFFECTS PRODUCED BY QUANTA

§ 43. *Nuclear scattering and absorption.* In the interaction of low energy (visible and ultra-violet) radiation with matter two distinct modes have long been recognised: absorption may take place for radiation of all frequencies greater than a critical frequency, or it may be selective. Then the frequency spectrum of the absorbed radiation is a line-spectrum. In the former event (photoelectric absorption), for each quantum of radiation absorbed, an electron is emitted from the atom which takes up the energy of the radiation; in the latter it is usual to say that the absorbing atom passes from one stationary state to another by the transition of an "optical" electron from a lower to a higher level of energy. With the investigation of radiation of higher frequencies (X-rays) a third type of interaction was revealed: a quantum (of sufficiently high frequency radiation) may be inelastically scattered by an atomic electron to which it thereby communicates energy. It was natural to suppose that with radiation of still higher frequency, that is with hard γ rays, similar effects might be observed which would have to be ascribed to interaction, possibly in all three modes, with the constituent particles of complex nuclei. In actual fact, however, the only true nuclear counterpart of the above effects at present recognised is the analogue of the first mentioned, photoelectric absorption: it will be discussed further in the next section. Here we must examine first the situation in respect of the other two.

In 1929 Kuhn[a] carried out an experiment by which it was hoped to show the greater absorption in ordinary lead than in uranium lead of some component of the γ radiation of thorium C″ (more particularly, of either or both of the components for which $E_\gamma = 2{\cdot}62$ and $0{\cdot}58 \times 10^6$ electron volts, respectively).

[a] Kuhn, *Phil. Mag.* 8, 625, 1929.

This appeared a possible result on the basis of the assumption that one or other (or both) of these components of the γ radiation (originating in nuclei of thorium lead left excited after the β particle disintegration Th C″→Th Pb) in fact arose in a transition to the ground state of this nucleus.[a] If that were so the transition in the opposite direction (absorption) should also occur, although it was realised that conditions in practice were far from ideally suited to testing this hypothesis. Doppler effects, due to thermal motion and radioactive recoïl (both β and γ), must spoil such an experiment completely if it so happens that the quantum radiation possesses a high degree of homogeneity. Kuhn's experiment eventually gave a negative result; naturally he concluded that the "half breadth of the γ ray line" was appropriately small (p. 145). Clearly, if a successful demonstration of the selective absorption of γ radiation by nuclei is to be effected, a more suitable choice of experimental conditions will have to be made.

The second phenomenon to which reference must be made in this section has been the subject of a large number of investigations in recent years. In 1930 Tarrant[b] and Chao[c] examined the absorption of the high energy γ radiation from thorium C″ in a number of substances. Within the limits of accuracy of their experiments they found that the electronic absorption coefficient[d] for this radiation, in light elements such as carbon and aluminium, was precisely that predicted on the basis of accepted theories. For heavier elements, however, an additional unexplained absorption process appeared to occur, which it was possible to characterise by an electronic coefficient proportional to the atomic number (an atomic

[a] It is now known that one, only, of these components arises in this way: cf. Oppenheimer (F.), *Proc. Camb. Phil. Soc.* 32, 328, 1936.

[b] Tarrant, *Proc. Roy. Soc.* 128, 345, 1930.

[c] Chao, *Proc. Nat. Acad. Sci.* 16, 431, 1930.

[d] The electronic absorption coefficient represents the fractional decrease in intensity, per extranuclear electron per sq. cm. area of a thin absorbing foil, which occurs when the radiation passes normally through the foil. It is also, numerically, the mean cross-sectional area per electron effective in the process of absorption. The atomic absorption coefficient is defined in a precisely similar manner. It may be regarded, when this point of view is significant, as the target area per atomic *nucleus* effective in absorption.

absorption coefficient, μ_a, proportional to Z^2). Rather naturally the tendency was to refer this additional absorption to the interaction of radiation with the nucleus—and the first attempt at a more formal explanation also followed these lines.[a] Now, however, the extra absorption is recognised as much less specifically nuclear in character. A detailed examination of the "additional" components of the scattered radiation[b] (as distinct from a direct investigation of the magnitude of the absorption) and expansion chamber experiments which established the production of pairs of positive and negative electrons by high frequency γ radiation (see § 27) led to this conclusion. The facts—and the interpretation generally accepted for them—are as follows.[c] Of the energy "anomalously" absorbed not all is re-radiated (scattered) in the form of a quantum radiation. From all elements that portion which is re-radiated consists largely of a component of about $0{\cdot}5 \times 10^6$ electron volts energy emitted isotropically, with radiation of roughly twice this energy also present in appreciable amount when heavy elements are employed. The production of positron-electron pairs varies with atomic number in the same way as the additional effects established by measurements of absorption and scattering. Other experiments[d] show the emission of radiation of $0{\cdot}5 \times 10^6$ electron volts energy when positive electrons are absorbed in matter—and $0{\cdot}5 \times 10^6$ electron volts is very closely the energy equivalent of the rest-mass of the electron. Empirically, therefore, there is every reason to attribute "additional absorption" to the process of pair production and additional scattering to the subsequent radiation from positive electrons. In the absorption process a quantum of energy E_γ disappears and a pair of electrons of total kinetic energy E is produced. If m_0 is the rest mass of each electron,

$$E_\gamma = 2m_0c^2 + E \qquad \ldots\ldots(64).$$

[a] Beck, *Naturwiss.* 18, 896, 1930.

[b] For work prior to 1934 see Gray and Tarrant, *Proc. Roy. Soc.* 143, 681, 706, 1934; for later work, Gentner, *Z. Physik*, 100, 445, 1936.

[c] Cf. Feather, *Ann. Rep. Chem. Soc.* 31, 368, 1935.

[d] Thibaud, *Phys. Rev.* 45, 781, 1934; Williams (E. J.), *Nature*, 133, 415, 1934; Klemperer, *Proc. Camb. Phil. Soc.* 30, 347, 1934.

Generally when most of the energy E has been dissipated in the production of ionisation, a favourable collision between the positive electron and a "free" negative electron takes place and energy in amount $2m_0c^2$ is again transformed into the energy of quanta. Two equal quanta are emitted and momentum is conserved. In this explanation everything has been included except the occasional production of "scattered" radiation of roughly 10^6 electron volts energy ($\sim 2m_0c^2$). Fermi and Uhlenbeck[a] first made the suggestion that this amount of available mass-energy might be radiated as a single quantum in a favourable collision of the positron with an electron tightly bound in a heavy atom.

It is interesting to see what theoretical justification there may be for this entirely empirical interpretation. In 1930 there was already in existence a theory due to Dirac which contained by implication the suggestion that pairs of oppositely charged particles might be "created" at the expense of the energy of radiation and, in other circumstances, "annihilate" one another with the production of radiation. When the relevance of these ideas to experiment was realised,[b] detailed calculations based upon them were found to predict for the process of creation a cross-section per nucleus[c] proportional to the square of the atomic number—with a dependence upon radiation frequency in good agreement with that found in respect of the additional absorption. Moreover, the absolute magnitude of the effect calculated was in close accord with that found by experiment.[d] Calculations concerning the process of annihilation have met with a similar success.[e] Here it is necessary to distinguish between annihilation when the positron has been reduced effectively to rest and annihilation when it still possesses considerable kinetic energy. The latter process is found to in-

[a] Fermi and Uhlenbeck, *Phys. Rev.* 44, 510, 1933; cf. Williams (E. J.), *Nature*, 135, 266, 1935.

[b] Cf. Blackett and Occhialini, *Proc. Roy. Soc.* 139, 699, 1933; Beck, *Z. Physik*, 83, 498, 1933.

[c] The process may be thought of as taking place in the intense electric field of the nucleus.

[d] Jaeger and Hulme, *Proc. Roy. Soc.* 153, 443, 1936.

[e] Dirac, *Proc. Camb. Phil. Soc.* 26, 361, 1930; Bhabha and Hulme, *Proc. Roy. Soc.* 146, 723, 1934; Bethe, *ibid.* 150, 129, 1935.

crease in probability, with respect to the former, both with the initial energy of the positron and (though more slowly) with the atomic number of the absorbing material. It may result in the production of one quantum or of two (unequal) quanta of radiation: again the partial probability of the one quantum process is found to increase very rapidly as the atomic number of the absorbing material increases. All these deductions from the theory will be seen to be in excellent accord with the empirical interpretation which has already been given.

§ 44. *Photo-disintegration.* The additional absorption of hard γ rays, discussed in the last section, is a process which varies regularly from element to element: no case of specific interaction was discovered until 1934. In that year, however, Chadwick and Goldhaber[a] showed that protons are liberated when a volume of heavy hydrogen is irradiated with the γ rays from a radiothorium source. They also found a positive effect, but one of considerably smaller magnitude (for equal γ ray intensity), when a radon source was employed. On the other hand, no protons were recorded in either case when the heavy hydrogen was replaced by ordinary hydrogen. Later,[b] they[c] reported the detection of neutrons in just those circumstances in which protons had previously been found. It was clear that, above a certain energy threshold, Q_γ, intermediate between the quantum energies of the intense high frequency components of the γ radiation from radium C and thorium C″, disintegration of the deuteron was taking place with the absorption of radiation. We may write

$$ {}_1\mathrm{H}^2 + Q_\gamma = {}_1\mathrm{H}^1 + {}_0n^1 \qquad \text{......(65)}, $$

$$ (1{\cdot}8 < Q_\gamma < 2{\cdot}6) \times 10^6 \text{ electron volts} \qquad \text{......(66)}. $$

An estimate of the amount of ionisation produced by the protons emitted when thorium C″ γ rays were employed suggested a value of about $2{\cdot}15 \times 10^6$ electron volts for Q_γ, consistent with (66). Shortly after the original experiments of

[a] Chadwick and Goldhaber, *Nature*, 134, 237, 1934.
[b] Chadwick and Goldhaber, *Proc. Roy. Soc.* 151, 479, 1935.
[c] See, also, Banks, Chalmers (T. A.) and Hopwood, *Nature*, 135, 99, 1935.

Chadwick and Goldhaber, Szilard and Chalmers[a] reported the photo-disintegration of beryllium—produced by the γ radiation from a sealed radium source, and detected by the radioactivity induced in iodine by the neutrons which were liberated. Recently, this transformation has been widely studied: it was of interest to decide what the charged residue might be, whether Be^8 or two α particles, and it was equally important to investigate the energy dependence of the disintegration yield. Here the first question concerns the γ ray threshold—and this was a matter of dispute for some time. Now the investigations of Brasch, Banks and others[b] and of Arzimowitsch and Palibin,[c] which set the limit between 1·35 and $1{\cdot}9 \times 10^6$ electron volts, have been confirmed by the more recent work of Chadwick and Goldhaber.[d] The latter authors give $Q_\gamma = 1{\cdot}6 \times 10^6$ electron volts, and with Bernardini and Mando[e] interpret this as the energy threshold for the two-particle process

$$ {}_4Be^9 + Q_\gamma = {}_4Be^8 + {}_0n^1 \qquad \ldots\ldots(67), $$

rather than for the three-particle transformation

$$ {}_4Be^9 + Q_\gamma' = 2\,{}_2He^4 + {}_0n^1 \qquad \ldots\ldots(68). $$

They are led to this conclusion both because of their failure to detect charged particles of more than about a millimetre range and also because no "slow" neutrons were observed.

Theoretically the photo-disintegration of the deuteron has been discussed from several points of view. On thermodynamical grounds it may be related to the reverse process, the combination of neutron and proton with the emission of radiation; with definite assumptions regarding the nature of the proton-neutron interaction the absolute probability of disintegration may also be calculated. These investigations have been carried out by Massey and Mohr[f] and by Bethe and Peierls,[g] amongst others. Fermi[h] was the first to propose

[a] Szilard and Chalmers, *Nature*, 134, 494, 1934.

[b] Brasch, Lange, Waly, Banks, Chalmers, Szilard and Hopwood, *Nature*, 134, 880, 1934.

[c] Arzimowitsch and Palibin, *Phys. Z. Sow. Un.* 7, 245, 1935.

[d] Chadwick and Goldhaber, *loc. cit.*

[e] Bernardini and Mando, *Phys. Rev.* 48, 468, 1935.

[f] Massey and Mohr, *Nature*, 133, 211, 1934; *Proc. Roy. Soc.* 148, 206, 1935.

[g] Bethe and Peierls, *Proc. Roy. Soc.* 148, 146, 1935.

[h] Fermi, *Phys. Rev.* 48, 570, 1935.

any important modification of the earlier calculations. He pointed out that, for the description of the radiation process involved in the combination of neutron and proton, a radiation field with both magnetic and electric dipole components was necessary. Previously the latter component only had been considered. When magnetic dipole radiation is included, obviously a somewhat greater cross-section for radiative capture (and so, also, for photo-disintegration) is predicted. Furthermore, the theoretical dependence of disintegration yield on quantum energy and the corresponding angular distribution of the disintegration particles are considerably modified.[a] With these changes such experiments as those of Lea[b] and of Bjerge and Westcott,[c] concerning the absorption of slow neutrons in hydrogenous substances, no longer present exceptional difficulties of interpretation. Moreover, experiment and theory remain in substantial agreement as concerns the photo-disintegration of heavy hydrogen by the γ rays of thorium C″.

The latest experiments here are those of Chadwick, Feather and Bretscher.[d] In them the energy threshold has been more accurately determined and the angular distribution of the disintegration products investigated. A mixture of helium and heavy methane was employed in a cloud chamber, irradiated by the γ rays from a radiothorium source. Plate III (*g*) is typical of the photographs of proton tracks so obtained. The mean length of track enabled Q_γ to be calculated, whilst the angular distribution was deduced directly from the observations. Subject to small corrections due to errors in the range-energy relation for slow protons, $Q_\gamma = 2{\cdot}26 \times 10^6$ electron volts: the experimental angular distribution was not inconsistent with the predominence of the electric dipole effect under the conditions of the experiment ($E_\gamma = 2{\cdot}62 \times 10^6$ electron volts).

With beryllium preliminary experiments of Chadwick and Goldhaber indicate quite a different angular distribution for

[a] Cf. Bethe and Bacher, *Rev. Mod. Phys.* 8, 82 (§§ 16, 17), 1936; Breit and Condon, *Phys. Rev.* 49, 904, 1936.

[b] Lea, *Nature*, 133, 24, 1934.

[c] Bjerge and Westcott, *Proc. Roy. Soc.* 150, 709, 1935.

[d] Chadwick, Feather and Bretscher, *Proc. Roy. Soc.* in course of publication.

the neutrons produced by photo-disintegration, and the recent work of Fleischmann and Gentner[a] confirms a rather surprising variation of disintegration yield with quantum energy. It will be interesting to see how far these differences can be explained in terms of the different quantum specifications of the ground states of the nuclei H^2 and Be^9.

So far photo-disintegration of light elements other than beryllium and deuterium has been looked for without success. This is not surprising if the masses of Table 6 be accepted as correct: with the γ rays of thorium active deposit no other case of photo-disintegration appears energetically possible—except the transformation

$$_3Li^6 \rightarrow {}_2He^4 + {}_1H^2 \qquad \ldots\ldots(69).$$

However, it would appear that the probability of this transformation is particularly small. It may be that this should be related to the vanishing electric dipole moment in the reverse process, the radiative capture of a deuteron by an α particle,[b] although magnetic dipole radiation may be found to be important here also.

With elements of medium and high atomic number information is rapidly accumulating concerning the radiation emitted when slow neutrons are captured;[c] this suggests that photo-disintegration (with the emission of neutrons) should become generally operative when γ rays of 10^7 electron volts energy are available. At present the penetrating radiation constitutes the only possible source; it is known[d] that quanta of roughly 10^7 electron volts energy are present at sea-level in appreciable numbers. Possibly evidence for photo-disintegration produced by this radiation will not long continue to be lacking.

[a] Fleischmann and Gentner, *Z. Physik*, 100, 440, 1936.
[b] Cf. Massey and Mohr, *loc. cit.*
[c] Kikuchi, Husimi and Aoki, *Nature*, 137, 992, 1936.
[d] Anderson, Millikan, Neddermeyer and Pickering, *Phys. Rev.* 45, 352, 1934.

NAME INDEX

NAME INDEX

NAME INDEX

NAME INDEX

NAME INDEX

Printed in the United States
By Bookmasters